# A Course in Linear Algebra

# A Course in Linear Algebra

GARETH WILLIAMS

*Department of Mathematics*
*University of Denver*

GORDON AND BREACH SCIENCE PUBLISHERS
New York London Paris

Gordon and Breach, Science Publishers, Inc.
440 Park Avenue South
New York, N.Y. 10016

*Editorial office for the United Kingdom*

Gordon and Breach, Science Publishers Ltd.
42 William IV Street
London W.C.2

*Editorial office for France*

Gordon & Breach
7–9 rue Emile Dubois
Paris 14^e^

Library of Congress catalog card number 73–156082. ISBN 0 677 03570 5.  Printed in east Germany.

To

BRIAN

# Preface

This book has arisen out of a series of lectures on linear algebra that I have given over a number of years. The undergraduate program of a mathematics major usually culminates with courses in linear algebra, algebraic structures and real analysis. Whichever of these courses the student takes first, it will, in all probability, be his first real exposure to abstract mathematical structures. It is my opinion that of these three courses, the student should be advised to take the linear algebra course first. The concepts of linear algebra are somewhat easier to grasp than those of the other fields and can be related to the properties of classical vectors which most students meet in the basic calculus courses. In fact, many students are taught a certain amount of linear algebra in the calculus sequence, in some cases relating the theory to a discussion of ordinary differential equations. This book assumes that the student is taking his first course in abstract mathematical structures. It is advantageous if the student has had some linear algebra in the calculus sequence, but not necessary.

I have included in the first six chapters topics that I consider essential in such a course. It should be possible to cover this material in a one quarter four hour course or a one semester three hour course. With time remaining in the course, the instructor can center upon a discussion of tensor algebra or of the spectral theory of operators, whichever he prefers, since the chapters presenting these topics are independent. These two chapters move more rapidly than the previous ones, it being assumed at this stage that the student has acquired a certain amount of maturity in the subject. These chapters are also suitable for graduate courses in linear algebra.

In addition to mathematicians, most scientists and engineers are finding that a knowledge of linear algebra is becoming necessary for their work. It is becoming increasingly important in the behavioral, social and management sciences. Linear algebra is the main area of mathematics used in the field of factor analysis, applied chiefly in psychology, but in recent years being used in economics, biology, physiology and medicine. It is the basis for linear programming and computational linear algebra, used in computer work. This book is intended for use as a text not only by mathematics students, but also by students from these other areas. Therefore, their needs are taken into account. For example, the application of the theory to differential equations is discussed at the appropriate places. The chapter on tensor algebra is of importance not only to mathematicians as background for differential geometry and manifold theory, but to scientists interested in such fields as relativity and fluid mechanics. The topics discussed in the chapter on spectral theory prepare the mathematics student for functional analysis; they also provide the theoretical physicist and theoretical chemist with much of the mathematical background necessary for quantum mechanics and give the engineer the algebra he needs for a qualitative discussion of differential equations. The psychologist engaged in applying factor analysis in testing programs will find the chapters on matrix representations and spectral theory of special interest.

I would like to thank my colleagues, Professors Herbert J. Greenberg, Frank G. Hagin, David L. Hector, Jean-Paul Marchand and Donald J. Rose, and my student, Mary S. Krimmel, for their helpful suggestions in the preparation of this text. I am also grateful to my wife Donna who criticized, proofread and typed much of the original manuscript, and to Laurie Lauridsen and Kathleen Russell for their part of the typing.

G. W.

# Contents

CHAPTER 1

# Finite Dimensional Vector Spaces

## 1.1 Set Theory

In this section we give the language that will be used to present linear algebra. Most of modern mathematics can be formulated using set theory; here some of the concepts and methods from the theory relevant to linear algebra will be given.

An *element* is an object of some sort such as a positive integer or a vector.

A *set* is a collection of elements. The statement "*x is an element of the set A*" is written $x \in A$ and "*x is not an element of A*" is written $x \notin A$.

A set is known once all its members are known. One way of indicating a set is to list its members, for example $\{1, 2, 3\}$ is a set consisting of the three elements 1, 2 and 3. However if there are many elements in the set this method is not very practical. The method of expressing a set then is to give, if possible, a membership condition; that is a criterion which tells us whether any given element is in the set or not. The notation used is $\{x: C(x)\}$ where $C(x)$ is symbolic for the condition that $x$ be an element of the set. For example the set $\{x: x = 2^n, n \text{ a positive integer}\}$ is the set consisting of the elements 2, 4, 8 etc.

Two sets $A$ and $B$ are said to be *equal* if they consist of exactly the same elements and we then write $A = B$. The negation of this statement is $A \neq B$. $A$ is said to be a *subset* of $B$ if each element of $A$ is also an element of $B$. This relation is written $A \subseteq B$. If $A$ is a subset of $B$ and not equal to $B$ it is said to be a *proper subset* of $B$ and we write $A \subset B$.

The symbol $\Rightarrow$ is used for logical implication. If $p$ and $q$ are statements, then $p \Rightarrow q$ means that $p$ implies $q$. Its negation is $p \not\Rightarrow q$. When the two statements are equivalent we will write $p \Leftrightarrow q$.

The following theorems give the main properties of set inclusion; their proofs illustrate the techniques used in set theory.

*Theorem 1*

a) $A \subseteq A$ for every set $A$. This is called the reflexive property of set inclusion.

b) $A \subseteq B$ and $B \subseteq A \Rightarrow A = B$. This is the anti symmetry property of set inclusion.

c) $A \subseteq B$ and $B \subseteq C \Rightarrow A \subseteq C$. Here we have the transitive property of set inclusion.

Proof: a) Let $xa \varepsilon A$, then trivially this implies that $x \varepsilon A$ and by the definition of subset, $A \subseteq A$.

b) Let $x \varepsilon A$, then $A \subseteq B \Rightarrow x \varepsilon B$.

If $y \varepsilon B$, then $B \subseteq A \Rightarrow y \varepsilon A$.

Hence $A$ and $B$ consist of the same elements and by the definition of equality of sets, $A = B$.

c) If $x \varepsilon A$, then since $A \subseteq B$, $x \varepsilon B$. Further since $B \subseteq C$, $x \varepsilon C$, and every element of $A$ is an element of $C$, implying that $A \subseteq C$.

*Theorem 2*

$A = B \Leftrightarrow A \subseteq B$ and $B \subseteq A$.

Proof: There are two parts to the proof of this theorem. First we have to prove that $A = B \Rightarrow A \subseteq B$ and $B \subseteq A$.

If $x \varepsilon A$ then since $A = B$, $x \varepsilon B$ and thus $A \subseteq B$. Similarly if $y \varepsilon B$ then since $A = B$, $y \varepsilon A$ and thus $B \subseteq A$. This proves the first part of the equivalence.

It remains to prove that $A \subseteq B$ and $B \subseteq A \Rightarrow A = B$. This is the result of theorem 1 b).

This theorem contains a useful method for proving the equality of two sets, namely to show that each is a subset of the other.

The *union* of two sets $A$ and $B$, written $A \cup B$, is the set of all elements which are in either $A$ or $B$ or in both. For example $\{1, 2, 3\} \cup \{2, 4, 6\}$ is the set $\{1, 2, 3, 4, 6\}$. The *intersection* of two sets $A$ and $B$, written $A \cap B$, is the set of all elements which are in both $A$ and $B$. $\{1, 2, 3\} \cap \{2, 4, 6\}$ is the set $\{2\}$. If $A$ and $B$ do not intersect they are said to be *disjoint*. The set consisting of no elements is called the *empty set* and is denoted by the symbol $\phi$. If $A$ and $B$ are disjoint then $A \cap B = \phi$.

We shall use diagrams extensively. The diagrams in themselves do not constitute proofs but should be of considerable aid in understanding the material presented. The concepts introduced thus far can be illustrated:

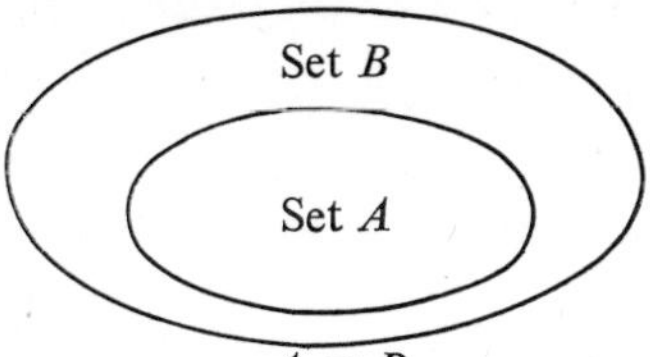


$A \subset B$

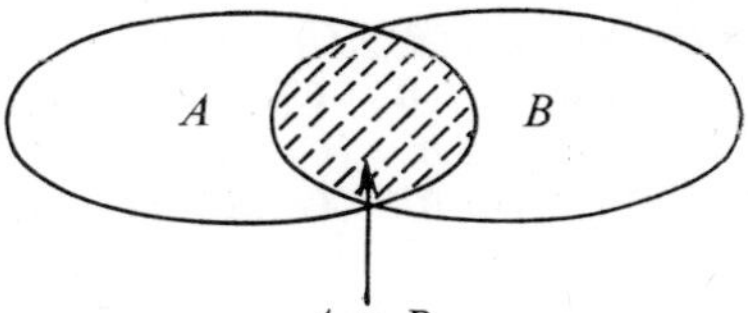


$A \cap B$

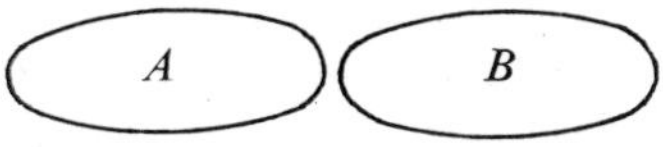


$A \cap B = \phi$

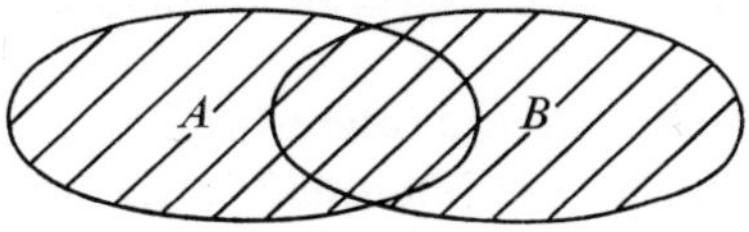


$A \cup B$

We now give an example which illustrates the application of theorem 2 in proving that two sets are equal.

*Example*

Prove that if $A \subset B$ then $A \cap B = A$.

Let $x \varepsilon A \cap B$. Then $x \varepsilon A$, implying that $(A \cap B) \subseteq A$. If $y \varepsilon A$, then $y \varepsilon B$ since $A \subset B$. Hence $y \varepsilon A \cap B$, implying that $A \subseteq (A \cap B)$.

Therefore $A \cap B = A$.

## Exercises

1 Identify the elements in the following sets of real numbers:
   a) $\{x: x^2 = 4\}$.
   b) $\{x: x^2 = -1\}$.
   c) $\{x: x + x = 0\}$.
   d) $A \cap B$, where $A = \{x: (x + 2)(x - 3) = 0\}$ and $B = \{x: (x + 2)^2 = 0\}$.
   e) $A \cup B$, where $A = \{x: (x + 2)(x + 3) = 0\}$ and $B = \{x: (x - 2)^2 = 0\}$.

2 List all the possible subsets of each of the following sets:
   a) $\{a\}$; b) $\{a, b, c\}$; c) $\phi$.

3 Give examples of two empty sets.

4 a) If $A$ is the set of all tennis players and $B$ is the set of all left-handed people describe $A \cap B$ and $A \cup B$.
   b) If $A$ is the set of days in April and $B$ is the set of days in May describe $A \cap B$ and $A \cup B$.

5 Verify:
   a) $\{x: x^2 + x + 1 = 0\} \subset \{x: x^3 = 1\}$.
   b) $\{x: x^3 - 8 = 0\} = \{x: (x - 2)(x^2 + 2x + 4) = 0\}$.
   c) $\{x: x^2 - 4 = 0\} \cap \{x: x^2 - 9\} = \phi$.
   The elements of all the sets are real numbers.

6 Prove that $A \subseteq B \Leftrightarrow A \cap B = A$.
7 Prove that $A \cap (B \cup C) = (A \cap B) \cup (A \cap C)$.
8 Prove that $A \cap (B \cap C) = (A \cup B) \cap (A \cup C)$.
9 Prove that $A \cap B = A \Rightarrow A \subset B$.
10 Prove that $A \cup B = A \cap B \Leftrightarrow A = B$.

## 1.2 Sets with Algebraic Structure

In algebra we are interested not only in collections of elements in sets and subsets; we are interested in sets and subsets which have rules for combining elements to give other elements. Here rules that we shall be using are given and their properties that are of interest defined.

A *binary operation* $*$ on a set is a rule that assigns to each ordered pair of elements $a$ and $b$ of the set a uniquely defined result object $a * b$, which may or may not be an element of the original set. Multiplication on the set of real numbers is an example of a binary operation. All the operations that we shall study will be binary operations so we shall call them operations and drop the term binary. A set with at least one operation is said to be a set with *algebraic structure*.

A set is *closed* with respect to an operation if for any two elements the result element is always a member of the set. The set of real numbers is closed under the operation of multiplication but the set of position vectors in Euclidean three-dimensional space is not closed under the operation of taking the dot product. In the latter case the result element is a real number, not a vector.

A set is *associative* under an operation $*$ if for any three elements $a$, $b$, and $c$ of the set

$$a * (b * c) = (a * b) * c.$$

That is, it does not matter which operation is performed first, $b * c$ or $a * b$. The positive integers are associative under the operation of addition. For example, $2 + (3 + 4) = (2 + 3) + 4$.

A set has the *commutative* property relative to an operation $*$ if for every pair of elements $a$ and $b$, $b * a = a * b$. Here again the set of positive integers under the operation of addition is an example of a set with this property. For example, $3 + 5 = 5 + 3$.

A set has an *identity* $e$ with respect to an operation $*$ if for an arbitrary element $a$ of the set, $a * e = e * a = a$. For the set of real numbers under the operation of addition the identity exists and is 0, for $a + 0 = 0 + a = a$ for any real number $a$. For the same set under the operation of multiplication the identity exists and is the real number 1, for $a \times 1 = 1 \times a = a$ for any real number $a$.

A set of elements has the *inverse property* with respect to an operation $*$ if the set has an identity $e$ with respect to the operation and if for every element $a$ of the set there exists an element $a'$ of the set such that $a * a' = a' * a = e$. The element $a'$ is called the inverse of $a$. Under the operation of addition the set of integers has the inverse property. For example the inverse of 3 is $-3$, since $3 + (-3) = (-3) + 3 = 0$. Under the operation of multiplication this set does not have the inverse property, for there does not exist an integer $a'$ such that $3 \times a' = a' \times 3 = 1$, demonstrating that 3 has no inverse under multiplication.

A *group* is a set of elements with an operation $*$ that has the following properties:

1 It is closed under the operation,
2 It is associative under the operation,
3 It has an identity element under the operation,
4 It has the inverse property under the operation.

As an example of a group we have the set of integers under the operation of addition. This set does not form a group under the operation of multiplication since, as we have seen, it does not have the inverse property.

A group is said to be an *Abelian* or *commutative group* if it is a group which is also commutative under the group operation. The set of position vectors in a three-dimensional Euclidean space forms an Abelian group under the operation of vector addition. Here the identity element is the zero vector, (0, 0, 0). The student is asked to prove this result in the exercises.

A set of elements with two operations $*$ and $\odot$ has the *distributive* property for $\odot$ with respect to $*$ if for every three elements $a$, $b$ and $c$ of the set

$$a \odot (b * c) = (a \odot b) * (a \odot c),$$

and

$$(a * b) \odot c = (a \odot c) * (b \odot c).$$

If this is the case we say that $\odot$ on the set is *distributive* with respect to $*$.

As an example of this property consider the set of real numbers under the operations of addition and multiplication. Multiplication is distributive with respect to addition, for if $a$, $b$ and $c$ are any three real numbers,

$$a \times (b + c) = (a \times b) + (a \times c),$$

and

$$(a + b) \times c = (a \times c) + (b \times c).$$

A *field* is a set of at least two elements and two operations $*$ and $\odot$ such that:

1 The set forms an Abelian group under the operation $*$. The identity here is usually called the *additive identity*.

2 With the identity under $*$ excepted, it forms an Abelian group under $\odot$. The identity of this group is usually called the *multiplicative identity*.
3 $\odot$ on the set is distributive with respect to $*$.

The set of real numbers forms a field under the operations of addition and multiplication. The additive identity is the number 0 and the multiplicative identity is the number 1.

### Exercises

1 Prove that the set of real numbers under the operation of addition forms an Abelian group.
2 Prove that the set of real numbers under the operations of addition and multiplication forms a field.
3 Prove that the set of position vectors in a three-dimensional Euclidean space forms an Abelian group under the usual law of vector addition.
4 Does the set $\{-1, 0, 1\}$, consisting of the three elements $-1$, 0 and 1 only, form an Abelian group under the usual operation of multiplication?
5 Consider a set whose elements are sets, the elements being $\{a, b\}$, $\{a\}$, $\{b\}$, $\phi$. For such a set the union of two elements is defined. Prove that union is a binary operation on the set. Is the set closed under this operation? Associative? Commutative? Does it have an identity? The inverse property? Is the set a group under this operation?
6 Discuss the properties of the set in question 5 under the operation of intersection. Is either operation distributive on the set with respect to the other operation? Is the set, with these two operations, a field?

## 1.3 Vector Spaces

The classical approach to the concept of a vector is to discuss vectors by means of components, in a certain special frame of reference suitably adapted to the situation. For example, in a rectangular cartesian frame in three dimensions one has the unit vectors $i$, $j$ and $k$ with components $(1, 0, 0)$, $(0, 1, 0)$ and $(0, 0, 1)$ and any other vector in the space can be written $(a, b, c)$ where $a$, $b$ and $c$ are real numbers and are the components of the vector in the directions $i$, $j$ and $k$ respectively. In such a space the vectors have certain properties, for example two vectors can be added to give a vector in the space. That is, the set of vectors is closed under the operation of addition. The purpose of this section is to axiomatize the classical properties of vectors. We shall draw up a set of axioms satisfied by such vectors, and any set of elements that satisfy these axioms will be called a vector space. Any set of elements that forms a vector space will have certain algebraic properties in common with classical vectors; the elements of such a set will be called vectors.

A *vector space* (or as it is sometimes called, a *linear space*) consists of the following:

1 A set of elements, called *vectors*, which forms an Abelian group under an operation called *addition*, denoted by $+$. The identity element is called the *zero vector* and is denoted $O$.

2 A field $\mathbf{F}$ of elements, called *scalars*, and an operation called *scalar multiplication* of vectors by scalars that results in vectors.

This operation satisfies the following conditions:

1 $a(bx) = (ab)x$, for scalars $a$ and $b$ and vector $x$; multiplication by scalars is associative.

2 $a(x + y) = ax + ay$, for scalar $a$ and vectors $x$ and $y$; multiplication by scalars is distributive with respect to vector addition.

3 $(a + b)x = ax + bx$; multiplication by vectors is distributive with respect to scalar addition.

4 For the element of the scalar field which is the multiplicative identity, call it 1, $1x = x$ for every vector $x$.

The relation between a vector space $V$ and the underlying field of scalars $\mathbf{F}$ is usually described by saying that *$V$ is a vector space over* $\mathbf{F}$. If $\mathbf{F}$ is the field $\mathbf{R}$ of real numbers, $V$ is called a *real vector space*; similarly, if $\mathbf{F}$ is the field of complex numbers, $V$ is called a *complex vector space*. These are the most common types of vector spaces and they are the ones that we shall be interested in here. We use the letter $\mathbf{F}$ to denote the field. The results will then apply to both types of vector spaces.

*Example 1*

In this example we consider the set of position vectors in three-dimensional Euclidean space and prove that it does indeed form a real vector space under the usual laws of vector addition and scalar multiplication. Call this space $\mathbf{R}^3$. The operations of addition and scalar multiplication on this space are defined by

$$(a, b, c) + (p, q, r) = (a + p, b + q, c + r),$$

and

$$h(a, b, c) = (ha, hb, hc),$$

where $(a, b, c)$ and $(p, q, r)$ are two arbitrary elements of $\mathbf{R}^3$ and $h$ is an arbitrary scalar.

We already know from the exercises in the previous section that $\mathbf{R}^3$ is an Abelian group under the operation of addition. We also know that the set of real numbers forms a field. Hence it remains to demonstrate that the axioms relating the Abelian group to the underlying field of scalars are satisfied.

Let $x = (a, b, c)$ and $y = (p, q, r)$ be arbitrary elements of $\mathbf{R}^3$ and let $h$, $k$ be arbitrary scalars. Then

1 $$\begin{aligned} h(kx) &= h(k(a, b, c)) = h(ka, kb, kc) = (hka, hkb, hkc) \\ &= (hk)(a, b, c) = (hk)x. \end{aligned}$$

2 $$\begin{aligned} h(x + y) &= h((a, b, c) + (p, q, r)) = h((a + p, b + q, c + r)) \\ &= (h(a + p), h(b + q), h(c + r)) = (ha + hp, hb + hq, hc + hr) \\ &= (ha, hb, hc) + (hp, hq, hr) = h(a, b, c) + h(p, q, r) \\ &= hx + hy. \end{aligned}$$

3 $$\begin{aligned} (h + k)x &= (h + k)(a, b, c) = ((h + k)a, (h + k)b, (h + k)c) \\ &= (ha + ka, hb + kb, hc + kc) = (ha, hb, hc) + (ka, kb, kc) \\ &= h(a, b, c) + k(a, b, c) = hx + kx. \end{aligned}$$

4 The multiplicative identity of the scalar field is the real number 1, and we have

$$1x = 1(a, b, c) = (1a, 1b, 1c) = (a, b, c) = x.$$

Hence all the axioms for a vector space are satisfied. The student is advised to go carefully through each step of the proofs of 1) ... 4) above seeing how each step depends upon the definitions of the operations.

*Example 2*

Consider the set of all real-valued continuous functions on the interval [0, 1]. Denote this set $F$. This set is called a *function space*. If $f$ and $g$ are elements of $F$ define an operation of addition on $F$ *pointwise*. That is, $f + g$ is an element of $F$ defined by $(f + g)(x) = f(x) + g(x)$ for every $x$ element of [0, 1]. This defines $f + g$ as a real-valued function on [0, 1] since it tells us how it operates on every element of [0, 1]. Under this operation the set forms an Abelian group:

If $f$ and $g$ are continuous functions on the interval [0, 1] then we know that the function $f + g$ defined by $(f + g)(x) = f(x) + g(x)$ is continuous on this interval. Hence the set is closed.

If $f$, $g$ and $h$ are elements of $F$, and $x$ an arbitrary element of [0, 1], then

$$\begin{aligned} (f + (g + h))(x) &= f(x) + (g + h)(x) \\ &= f(x) + g(x) + h(x) \\ &= (f + g)(x) + h(x) \\ &= ((f + g) + h)(x), \end{aligned}$$

implying that

$$f + (g + h) = (f + g) + h.$$

The set is associative.

Let $O$ be the zero function, the function that takes all the elements of the set [0, 1] into zero. Then

$$(f + O)(x) = f(x) + O(x)$$
$$= f(x).$$

Similarly

$$(O + f)(x) = f(x).$$

This implies that

$$f + O = f = O + f.$$

$O$ is the identity element of $F$.

Let $-f$ be the continuous function on [0, 1] defined by

$$(-f)(x) = -f(x).$$

Then

$$(f + (-f))(x) = f(x) + (-f)(x)$$
$$= f(x) - f(x)$$
$$= 0$$
$$= O(x),$$

proving that

$$f + (-f) = O.$$

Similarly $(-f) + f = O$. Hence $(-f)$ is the inverse of $f$ and the set has the inverse property.

Finally,

$$(f + g)(x) = f(x) + g(x)$$
$$= g(x) + f(x)$$
$$= (g + f)(x),$$

implying that

$$f + g = g + f.$$

The set is commutative. It is thus an Abelian group.

Next define an operation of scalar multiplication of the elements of $F$ by real numbers. For $a$ being an arbitrary real number and $f$ an arbitrary element of $F$, let $af$ be the element of $F$ defined by $(af)(x) = af(x)$, for every $x$ element of [0, 1]. This is called *pointwise multiplication* of $f$ by the real number $a$.

Under these operations the set of all real-valued continuous functions on the interval [0, 1] is a real vector space. The student is asked to complete this proof in the exercises.

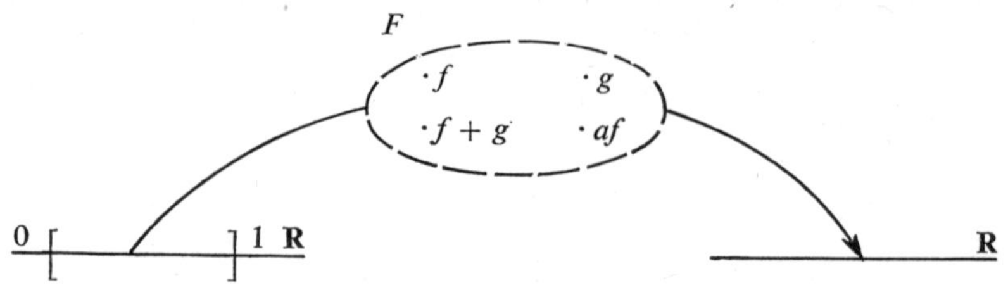


*Example 3*

The set of real numbers over the field of real numbers forms a real vector space. Here the set of real numbers is an Abelian group, hence its elements can be taken as vectors; it is also a field, so can be taken as the scalar field. The axioms giving the interaction between the Abelian group and the scalar field for a vector space can be seen to be satisfied.

In a similar manner the set of complex numbers over the field of complex numbers forms a complex vector space.

We shall come across further examples of vector spaces from time to time. This algebraic structure is the basic structure that we will be interested in throughout this book.

Certain subsets of vector spaces form vector spaces in their own right under the operations of the larger set. Such a subset is called a *subspace* of the vector space. A subset of a vector space is a subspace if and only if it is closed under the operations of addition and scalar multiplication. That is, if and only if, for $u$ and $v$ elements of the subset, $u + v$ is an element of the subset, and for any scalar $a$, $au$ is an element of the subset. If these conditions are satisfied by the subset all the other conditions of a vector space are satisfied from the properties of the larger set. Of course if either of these conditions does not hold the subset cannot be a vector space.

**Exercises**

1 Complete the proof of example 2 of this section.

2 Prove that the set of all differentiable real-valued functions on the interval [0, 1], with operations of addition and scalar multiplication as defined in the vector space of continuous real-valued functions, forms a real vector space.

3 Consider the set of all real-valued functions (not necessarily continuous) on the interval [0, 1]. Is this set a vector space under operations of pointwise addition and pointwise multiplication? If so, is the set of continuous real-valued functions of example 2 a subspace of this space? Is the set of all discontinuous real-valued functions, on the same interval, a vector space under the same two operations?

4 From example 1 we know that the set of position vectors in three-dimensional Euclidean space forms a vector space. Label these position vectors relative to the origin, using the usual $x$, $y$, $z$ coordinates. Prove that the position vectors in the $xy$-plane form a subspace of this vector space.

5 Prove that if $U$ is a subset of a vector space $V$ which is closed under the operations of addition and scalar multiplication which make $V$ into a vector space, then it is a vector space in its own right. It is called a *subspace* of $V$.

6 Let $V$ be a vector space over a field $F$ and $U$ a subset of $V$. Prove that a necessary and sufficient condition for $U$ to be a subspace of $V$ is that $au_1 + bu_2$ be an element of $U$ for arbitrary elements $u_1$ and $u_2$ of $U$ and arbitrary scalars $a$ and $b$. (We shall often use this condition to determine if a subset is a subspace.)

7 Let $V$ be a vector space, $o$ the additive identity of the scalar field (usually called the *zero scalar*) and $v$ an arbitrary element of $V$. Prove that $ov = O$, the zero vector.

## 1.4 The Vector Space $\mathbf{R}^n$

Let $A_1, \ldots, A_n$ be $n$ sets, $n$ being a positive integer. The *Cartesian product* of these sets, denoted $A_1 \times A_2 \times \cdots \times A_n$, is defined as the set

$$\{(x_1, x_2, \ldots, x_n) : x_1 \varepsilon A_1, \ldots, x_n \varepsilon A_n\}.$$

An element of this set $(x_1, x_2, \ldots, x_n)$ is said to be an *n-tuple.*

For example let $A_1$ be the set $\{x : a \le x \le b,\ x \text{ a real number}\}$, denoted $[a, b]$ and $A_2$ the set $[c, d]$. Then

$$A_1 \times A_2 = \{(x_1, x_2) : x_1 \varepsilon [a, b], x_2 \varepsilon [c, d]\}.$$

This set can be represented as the subset of the plane shaded in the following diagram.

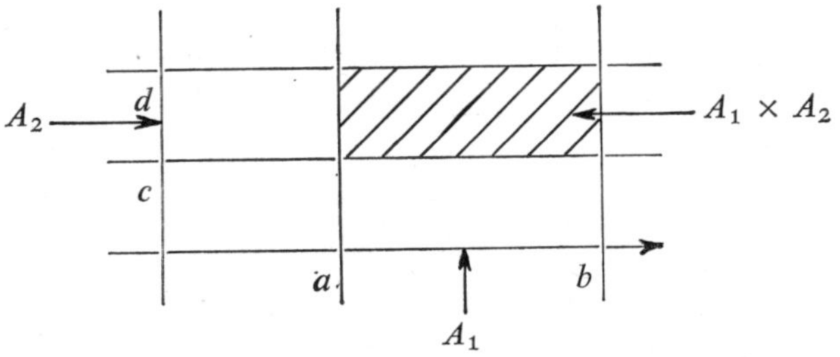


Let us denote the set of real numbers by $\mathbf{R}$. Then, as has been stated, this set over the field of real numbers forms a real vector space. Denote the set $\mathbf{R} \times \cdots \times \mathbf{R}$, $n$ times ($n$ a positive integer), by $\mathbf{R}^n$.

That is, $\mathbf{R}^n = \{(x_1, x_2, \ldots, x_n) : x_1, \ldots, x_n \varepsilon \mathbf{R}\}$.

An element of $\mathbf{R}^n$ is an $n$-tuple of real numbers and $\mathbf{R}^n$ is the set of all $n$-tuples of real numbers. Note that for $n > 1$, $\mathbf{R}^n$ as it has been defined above has no structure whatsoever. In different situations mathematicians put varying structures on $\mathbf{R}^n$. In this book we shall be interested in putting a vector-space structure on $\mathbf{R}^n$. This can be done in a natural way using the operations of $\mathbf{R}$ ($\mathbf{R}$ is a field).

Use the field operation of addition on **R** to define an operation of addition on $\mathbf{R}^n$ as follows:

For $(x_1, \ldots, x_n)$ and $(y_1, \ldots, y_n)$ elements of $\mathbf{R}^n$ let $(x_1, \ldots, x_n) + (y_1, \ldots, y_n) = (x_1 + y_1, \ldots, x_n + y_n)$. Under this operation the elements of $\mathbf{R}^n$ form an Abelian group. The identity element, $O$, of this group is $(0, \ldots, 0)$.

Define the field of underlying scalars to be **R** and let the operation of scalar multiplication be $y(x_1, \ldots, x_n) = (yx_1, \ldots, yx_n)$, for an arbitrary scalar $y$.

These can be seen to be generalizations of the operations that were defined on $\mathbf{R}^3$ in the previous section, making it into a vector space.
Under these two operations the set $\mathbf{R}^n$ becomes a vector space.

The reader is familiar with the set $\mathbf{R}^3$ as representing the position vectors of elements in three-dimensional Euclidean space, relative to the origin. It should be noted however that this space of position vectors does have further properties such as dot product which we have not defined on $\mathbf{R}^3$. This is an additional algebraic structure which we shall soon define rigorously.

Throughout this text $\mathbf{R}^n$ will be taken to be the real vector space defined here.

In a similar manner, if **F** is an arbitrary field and $\mathbf{F}^n = \mathbf{F} \times \cdots \times \mathbf{F}$, then $\mathbf{F}^n$ can be interpreted as a vector space over the field of scalars **F**. The elements of $\mathbf{F}^n$ can be interpreted as row vectors or as column vectors. For the row-vector interpretation, operations of addition and scalar multiplication are defined by

$$(a_1, \ldots, a_n) + (b_1, \ldots, b_n) = (a_1 + b_1, \ldots, a_n + b_n),$$
$$a(a_1, \ldots, a_n) = (aa_1, \ldots, aa_n);$$

and for the column-vector interpretation,

$$\begin{pmatrix} a_1 \\ \vdots \\ a_n \end{pmatrix} + \begin{pmatrix} b_1 \\ \vdots \\ b_n \end{pmatrix} = \begin{pmatrix} a_1 + b_1 \\ \vdots \\ a_n + b_n \end{pmatrix},$$

$$a\begin{pmatrix} a_1 \\ \vdots \\ a_n \end{pmatrix} = \begin{pmatrix} aa_1 \\ \vdots \\ aa_n \end{pmatrix}.$$

We shall see that in certain situations it is more convenient to think of the elements of $\mathbf{F}^n$ as column vectors than as row vectors.

## Exercises

1 Let $A = [1, 2]$, $B = [1, 3]$, $C = [2, 4]$. Define the set $A \times B \times C$ and illustrate with a diagram.

2 Define and illustrate with a diagram the set $([-2, 3] \times [0, 2]) \cap ([-1, 4] \times [1, 3])$.

3 Define and illustrate with a diagram the set $([0, 1] \times [0, 1]) \cup ([-1, 0] \times [-1, 0])$.
4 Prove that $\mathbf{R}^n$ with the operations of addition and scalar multiplication defined in this section is a vector space.
5 Prove that if $\mathbf{F}$ is a field, $\mathbf{F}^n$ can be made into a vector space over $\mathbf{F}$ by defining suitable operations of addition and scalar multiplication.
6 If $A = \{a, b, c, d\}$ and $B = \{a, c, b, q, r\}$, define $A \times B$.
7 Let $U$ and $V$ be non-empty sets. If $A_1$ and $A_2$ are subsets of $U$ and $B_1$ and $B_2$ are subsets of $V$, prove

a) $(A_1 \times B_1) \cap (A_2 \times B_1) = (A_1 \cap A_2) \times B_1$;

b) $(A_1 \times B_1) \cap (A_2 \times B_2) = (A_1 \cap A_2) \times (B_1 \cap B_2)$.

Illustrate these results with a diagram for $U$ and $V$ both being the set of real numbers and $A_1$, $A_2$, $B_1$, $B_2$ being subsets which can be represented by intervals of the real line.

## 1.5 Bases for Vector Spaces

Let $V$ be a vector space over a field of scalars $\mathbf{F}$. We say that $v \in V$ can be expressed as a *linear combination* of $v_1, \ldots, v_n$, (or of the set $\{v_1, \ldots, v_n\}$), for $v_1, \ldots, v_n$ $n$ vectors in $V$, if there exist $n$ scalars $a_1, \ldots, a_n$, such that

$$v = a_1 v_1 + \cdots + a_n v_n.$$

*Example 1*

In the space $\mathbf{R}^3$, the vector $(1, 2, 3)$ can be expressed as a linear combination of the four vectors $(1, 0, 0)$, $(0, 3, 0)$, $(0, 1, 0)$ and $(0, 0, 1)$. For,

$$(1, 2, 3) = 1(1, 0, 0) + 1(0, 3, 0) - 1(0, 1, 0) + 3(0, 0, 1).$$

The aim of this section is to demonstrate that there exist certain special subsets of $V$, called *bases*, such that all other vectors in $V$ can be expressed uniquely as linear combinations of the elements of any one of these subsets. In discussing operations on $V$ it will be shown that one then need only look at the operations on such a special subset and one will know what is happening on the whole space. To discuss this concept we need the following definitions.

In the vector space $V$ the set $\{v_1, \ldots, v_n\}$ is said to *span* $V$ or *generate* $V$ if every vector in $V$ can be expressed as a linear combination of the set.

*Example 2*

The set $\{(1, 0, 0), (0, 3, 0), (0, 1, 0), (0, 0, 1)\}$ spans $\mathbf{R}^3$ since any vector $(a, b, c)$ can be expressed as a linear combination of its elements:

$(a, b, c) = a(1, 0, 0) + p(0, 3, 0) + q(0, 1, 0) + c(0, 0, 1)$, where $3p + q = b$; $p$ and $q$ will not be unique. Note that the set $\{(1, 0, 0), (0, 1, 0), (0, 0, 1)\}$ also spans $\mathbf{R}^3$.

Hence a set of vectors that spans $V$ does characterize the space in that it can be used to generate any element of $V$. However a set that spans $V$ need not be the most efficient set for characterizing $V$, a proper subset could still span the space. We find the most efficient set, a basis, to be a set that spans $V$, the elements of which are in some sense independent. To formulate this concept we introduce linear dependence and independence of vectors.

The set $\{v_1, \ldots, v_n\}$ is said to be *linearly dependent* if there exist $n$ scalars $a_1, \ldots, a_n$, not all equal to zero, such that

$$a_1 v_1 + \cdots + a_n v_n = O.$$

The vectors are then said to be linearly dependent.

If $v_1, \ldots, v_n$ are not linearly dependent they are said to be *linearly independent*.

*Example 3*

The set of vectors $\{(2, -1, 0), (-4, 2, 0), (0, 0, 3)\}$ is linearly dependent in $\mathbf{R}^3$. To see this we examine the identity

$$a_1(2, -1, 0) + a_2(-4, 2, 0) + a_3(0, 0, 3) = O.$$

This identity reduces to

$$(2a_1, -a_1, 0) + (-4a_2, 2a_2, 0) + (0, 0, 3a_3) = O,$$
$$(2a_1 - 4a_2, -a_1 + 2a_2, 3a_3) = O.$$

Hence the scalars $a_1$, $a_2$ and $a_3$ must satisfy the system of linear equations

$$\begin{aligned} 2a_1 - 4a_2 &= 0, \\ -a_1 + 2a_2 &= 0, \\ 3a_3 &= 0. \end{aligned}$$

This system has solutions $a_1 = 2a_2$, $a_3 = 0$. $a_1 = 2$, $a_2 = 1$, $a_3 = 0$ satisfy these conditions and the vectors are thus linearly dependent.

*Example 4*

The vectors $(0, 0, 2)$, $(2, 1, 0)$ and $(-1, 2, 1)$ are linearly independent in $\mathbf{R}^3$. To see this we examine the identity

$$a_1(0, 0, 2) + a_2(2, 1, 0) + a_3(-1, 2, 1) = O.$$

This reduces to the system of equations

$$\begin{aligned} 2a_2 - a_3 &= 0, \\ a_2 + 2a_3 &= 0, \\ 2a_1 + a_3 &= 0, \end{aligned}$$

the only solution of which is easily seen to be $a_1 = a_2 = a_3 = 0$. Thus the scalars are zero and the vectors are linearly independent.

We are now in a position to define a basis for a vector space. The set $\{v_1, \ldots, v_n\}$ forms a *basis* for $V$ if the elements span $V$ and are linearly independent.

*Example 5*

The set $\{(0, 0, 2), (2, 1, 0), (-1, 2, 1)\}$ forms a basis for $\mathbf{R}^3$.

We saw in the previous example that this set is linearly independent. It also spans $\mathbf{R}^3$.

*Example 6*

Let $V$ be the real vector space generated by the two functions $\sin t$, $\cos t$ defined on the interval $[0, \pi]$. Then any element of $V$ is of the form $a \sin t + b \cos t$, $a, b \in \mathbf{R}$.

The identity $a \sin t + b \cos t = 0$ gives, on differentiation, $a \cos t - b \sin t = 0$. Solving these equations one finds that $a = b = 0$, implying that the vectors $\sin t$ and $\cos t$ are linearly independent. They also span $V$, by definition of the vector space, and therefore form a basis for $V$.

*Theorem 1*

If $\{v_1, \ldots, v_n\}$ is a basis for a vector space $V$ then each vector in $V$ can be expressed uniquely as a linear combination of these vectors.

Proof: By the definition of basis we know that each vector in $V$ can be expressed as a linear combination of $\{v_1, \ldots, v_n\}$. Hence it remains to prove the uniqueness.

Suppose the linear combination is not unique for some vector $v$, that is, that there exist two distinct sets of scalars $\{a_i\}$ and $\{b_i\}$ such that:

$$v = \sum_{i=1}^{n} a_i v_i \quad \text{and} \quad v = \sum_{j=1}^{n} b_j v_j.$$

Hence

$$\sum_{i=1}^{n} a_i v_i = \sum_{j=1}^{n} b_j v_j,$$

giving

$$\sum_{i=1}^{n} (a_i - b_i)\, v_i = O.$$

Since the vectors $v_1, \ldots, v_n$ form a basis and are therefore linearly independent, each coefficient in this last equation must be identically zero. Hence $a_i = b_i$, $i = 1, \ldots, n$. This contradicts the assumption that the two linear combinations are distinct. Hence this assumption is incorrect; each linear combination is unique.

*Theorem 2*

If $\{v_1, \ldots, v_n\}$ is a basis for a vector space $V$, then it is a *maximal set* of linearly independent vectors. That is, any set containing this set as a proper subset is a set of linearly dependent vectors.

Proof: Let $\{v_1, \ldots, v_n, v_{n+1}, \ldots, v_p\}$ be a set of vectors containing the set $\{v_1, \ldots, v_n\}$ as a proper subset.

Since $\{v_1, \ldots, v_n\}$ is a basis for $V$, the vector $v_{n+1}$ can be expressed as a linear combination of these vectors.

Let

$$v_{n+1} = a_1 v_1 + \cdots + a_n v_n.$$

Then,

$$a_1 v_1 + \cdots + a_n v_n - v_{n+1} + o v_{n+2} + \cdots + o v_p = O,$$

proving that $\{v_1, \ldots, v_p\}$ is a set of linearly dependent vectors.

*Example 7*

The set $\{(1, 0, 0), (0, 1, 0), (0, 0, 1)\}$ forms a basis for $\mathbf{R}^3$. Any vector in $\mathbf{R}^3$ can be expressed uniquely as a linear combination of these three vectors. For example,

$$(a, b, c) = a(1, 0, 0) + b(0, 1, 0) + c(0, 0, 1).$$

Any set of vectors in $\mathbf{R}^3$ containing these three vectors is a set of linearly dependent vectors. For example $\{(1, 2, 3), (1, 0, 0), (0, 1, 0), (0, 0, 1)\}$ is a set of linearly dependent vectors, for $1(1, 2, 3) - 1(1, 0, 0) - 2(0, 1, 0) - 3(0, 0, 1) = (0, 0, 0) = O$.

## Exercises

1 Prove that the set $\{(1, 0), (0, 1)\}$ forms a basis for $\mathbf{R}^2$. Can you find another set of vectors that forms a basis for $\mathbf{R}^2$?

2 Show that the set $\{(2, 3), (1, 2), (0, 1)\}$ is linearly dependent in $\mathbf{R}^2$.

3 Let $V$ be a vector space. Prove that any subset of $V$ that contains the zero vector is linearly dependent.

4 Let $V$ be a vector space and $\{v_1, \ldots, v_n\}$ a set of linearly dependent vectors. Prove that any set of vectors that contains this set as a subset is linearly dependent.

5 Prove the following:
$v_1, \ldots, v_n$ are linearly dependent $\Leftrightarrow$ one of the vectors can be expressed as a linear combination of the others.
$v_1, \ldots, v_n$ are linearly independent $\Leftrightarrow$ it is not possible to express one of the vectors as a linear combination of the others.

6 Let $1, x, x^2$ be elements of the vector space of differentiable functions on the interval $[0, 1]$ discussed in exercise 2 of section 1.3. Prove that these vectors are linearly independent. Define a subspace of the vector space of all differentiable functions on $[0, 1]$ generated by these vectors. Hint—use a method similar to that of example 6 to prove the linear independence of these vectors.

7 Prove that $e^x$ and $e^{2x}$ are linearly independent vectors in the vector space of real valued differentiable functions on the interval [0, 1]. Define a subspace of this vector space generated by $e^x$ and $e^{2x}$. Give an alternative basis for this space, justifying your result.

## 1.6 Dimension of a Vector Space

In the last section we developed the concept of basis and discussed how we could use the basis to represent the space. Here we develop further the concept of basis and prove that in the vector spaces of interest to us all bases for a particular vector space have the same number of elements.

*Theorem 1*

Let $V$ be a vector space with basis $\{v_1, \ldots, v_n\}$ and let $W = \{w_1, \ldots, w_m\}$ be a set of distinct vectors in $V$ with $m > n$. Then $w_1, \ldots, w_m$ are linearly dependent.

Proof: If $\{w_1, \ldots, w_m\}$ contains the zero vector it is linearly dependent (ex. 3, section 1.5). Hence it remains to consider the case when all the vectors are non-zero.

If there does not exist a subset of $n$ linearly independent vectors in $W$, then a subset of $W$ consisting of $n$ vectors is linearly dependent, hence by ex. 4, section 1.5, $W$ is linearly dependent. Therefore it remains to consider the case when there exists a subset $W'$ of $W$ containing $n$ lineary independent vectors. Re-label these vectors if necessary so that they are $w_1, \ldots, w_n$.

Since $\{v_1, \ldots, v_n\}$ is a basis, $w_1$ can be expressed as a linear combination of these elements.

Let $w_1 = \sum_{i=1}^{n} a_i v_i$.

Since $w_1$ is not the zero vector there exists at least one non-zero $a_i$. It can be assumed without loss of generality that it is $a_1$. Hence the equation can be arranged to read

$$v_1 = \frac{1}{a_1} w_1 - \frac{a_2}{a_1} v_2 \cdots - \frac{a_n}{a_1} v_n.$$

The space spanned by $w_1, v_2, \ldots, v_n$ contains $v_1$ and hence this set spans $V$. It therefore contains $w_2$. In the same way as above the vector $v_2$ can be replaced by $w_2$ and the set $\{w_1, w_2, v_3, \ldots, v_n\}$ spans $V$. Continuing thus it can be seen that the set $\{w_1, \ldots, w_n\}$ spans $V$. Since it is linearly independent it forms a basis. This is a proper subset of $W$, hence by theorem 2, section 1.5, $W$ is a set of linearly dependent vectors.

An alternative way of stating this theorem is to say that if $\{v_1, \ldots, v_p\}$ is a set of linearly independent vectors in a space of dimension $n$, then $p \leqq n$. This form of the theorem is now used to prove that the number of vectors that make up a basis is unique.

*Theorem 2*

The number of vectors that make up a basis of a vector space is unique, if this number is finite.

Proof: Let $\{v_1, \ldots, v_n\}$ and $\{w_1, \ldots, w_m\}$ be two bases for $V$. Then both sets are linearly independent. Hence by theorem 1, $m \leqq n$ and $n \leqq m$. Therefore $m = n$.

This theorem enables us to associate with every such vector space a positive integer called the *dimension* of the space. Let $V$ be a vector space having a basis consisting of $n$ vectors. Then $n$ is the dimension of this space. If the vector space consists of the zero vector only we say that it has *dimension zero*. Such vector spaces are said to be *finite-dimensional.*

*Theorem 3*

In a space of dimension $n$, any set of $n$ linearly independent vectors forms a basis.

Proof: Let $\{v_1, \ldots, v_n\}$ be a set of $n$ linearly independent vectors in a vector space of dimension $n$. Then for any other vector $w$ in the space the set $\{v_1, \ldots, v_n, w\}$ is linearly dependent by theorem 1. Hence $\{v_1, \ldots, v_n\}$ spans $V$ and is therefore a basis.

In this text we shall be primarily interested in vector spaces of finite dimensions. Spaces of infinite dimension do exist but many of the theorems that hold for spaces of finite dimensions do not carry over to those of infinite dimension. The function space discussed in example 2 of section 1.3 is an infinite-dimensional vector space. Such spaces are of importance and students who go on to do graduate work in mathematics will meet them in a course on functional analysis.[1]

*Example 1*

For $\mathbf{R}^3$ the set $\{(1, 0, 0), (0, 1, 0), (0, 0, 1)\}$ forms a basis. This is called the *usual basis* for $\mathbf{R}^3$. Hence the dimension of $\mathbf{R}^3$ is 3. The usual basis for $\mathbf{R}^n$ is defined to be the set $\{(1, 0, \ldots, 0), (0, 1, \ldots, 0), \ldots (0, \ldots, 0, 1)\}$ and this space is of dimension $n$.

*Theorem 4*

Let $V$ be a vector space of dimension $n$ and $\{v_1, \ldots, v_r\}$, $r < n$, be a set of $r$ linearly independent vectors in $V$. Then there exist vectors $v_{r+1}, \ldots, v_n$ in $V$ such that the set $\{v_1, \ldots, v_r, v_{r+1}, \ldots, v_n\}$ forms a basis for $V$. That is, a set of linearly independent vectors can always be extended to form a basis for the space.

---

[1] See, for example, K. Yosida, *Functional Analysis* (Springer, 1965); G. F. Simmons, *Topology and Modern Analysis* (McGraw-Hill, 1963).

Proof: The set $\{v_1, \ldots, v_r\}$ cannot span $V$, for otherwise the dimension of $V$ would be $r$ and not $n$. Hence there exists a vector $v_{r+1}$ in $V$ which is not a linear combination of $v_1, \ldots, v_r$. Then if $r + 1 = n$, the set $\{v_1, \ldots, v_{r+1}\}$ is a set of $n$ linearly independent vectors in $V$ and hence forms a basis for $V$. If $n > r + 1$, then a vector $v_{r+2}$ can be added to the set such that the set $\{v_1, \ldots, v_{r+2}\}$ is linearly independent. Continuing thus $i$ vectors can be added until $r + i = n$. Then the set $\{v_1, \ldots, v_r, v_{r+1}, \ldots, v_{r+i}\}$ forms a basis for $V$.

### Exercises

1 Find two distinct sets of base vectors for $\mathbf{R}^3$.

2 Give examples to illustrate

a) theorem 1, for $\mathbf{R}^3$

b) theorem 4, for $\mathbf{R}^3$

3 Extend the set $\{(1, 2, 3), (0, 1, 2)\}$ to form a basis for $\mathbf{R}^3$.

4 Extend the set $\{(1, 4, 0), (4, 1, 3)\}$ to form a basis for $\mathbf{R}^3$.

5 Prove that the subset of vectors $\{(x_1, x_2, x_3, x_4): x_1 - x_2 = 0\}$ in $\mathbf{R}^4$ is a subspace of $\mathbf{R}^4$. Find a basis for this subspace. What is its dimension?

6 Prove that the subset of vectors $\{(x_1, x_2, x_3, x_4): x_1 = x_2 = 0\}$ in $\mathbf{R}^4$ is a subspace of $\mathbf{R}^4$. Find a basis for this subspace and give its dimension.

7 a) Prove that if $V_1$ is a proper subspace of a finite-dimensional vector space $V_2$ then $\dim V_1 < \dim V_2$.

b) If $V_1$ is a subspace of a finite-dimensional vector space $V_2$ such that $\dim V_1 = \dim V_2$, prove that $V_1 = V_2$.

## 1.7 Intersections of Vector Spaces

*Theorem 1*

Let $V_1$ and $V_2$ be subspaces of a vector space $V$. Then $V_1 \cap V_2$ is a subspace of $V$.

Proof: For this result to hold it is necessary and sufficient to prove that $V_1 \cap V_2$ is closed under addition and scalar multiplication.

Let $u$ and $v$ be elements of $V_1 \cap V_2$. Then $u$ and $v$ are both elements of $V_1$ since $V_1 \cap V_2$ is a subset of $V_1$. Since $V_1$ is closed under addition $u + v$ is an element of $V_1$. Similarly $u + v$ is an element of $V_2$ and hence $u + v$ is an element of $V_1 \cap V_2$. $V_1 \cap V_2$ is therefore closed under addition.

Let $c$ be a scalar. $cv$ is an element of $V_1$ since $V_1$ is closed under scalar multiplication. Similarly $cv$ is an element of $V_2$. Hence $cv$ is an element of $V_1 \cap V_2$.

$V_1 \cap V_2$ is therefore a subspace of $V$.

Let $V_1$ and $V_2$ be subspaces of a vector space $V$. The *sum* of $V_1$ and $V_2$, denoted by $V_1 + V_2$, is the subset of $V$ consisting of all vectors of the form $v_1 + v_2$, where $v_1$ is an element of $V_1$ and $v_2$ is an element of $V_2$.

*Theorem 2*

Let $V$ be a vector space with subspaces $V_1$ and $V_2$. Then $V_1 + V_2$ is a subspace of $V$.

The reader is asked to prove this result in exercise 2.

We say that a vector space $U$ is the direct sum of $V_1$ and $V_2$, written $U = V_1 \oplus V_2$, if and only if $U$ is the sum of $V_1$ and $V_2$ and each vector $u$ of $U$ is expressible uniquely in the form $u = u_1 + u_2$ with $u_1$ an element of $V_1$ and $u_2$ an element of $V_2$. It follows from theorem 2 that $V_1 \oplus V_2$ is a subspace of $V$.

*Theorem 3*

Let $V$ be a vector space and $V_1$, $V_2$ subspaces. Then $V = V_1 \oplus V_2$ if and only if $V = V_1 + V_2$ and $V_1 \cap V_2 = \{O\}$. That is, the sum of two subspaces is a direct sum if and only if the intersection of the two subspaces consists of only the zero vector.

The proof, since it is of the if-and-only-if type, has two parts.

Proof: a) If $V = V_1 + V_2$ and $V_1 \cap V_2 = \{O\}$, it is necessary to prove that $V = V_1 \oplus V_2$.

Let $v \, \varepsilon \, V$. Suppose there exist elements $v_1$ and $v_1'$ in $V_1$ and $v_2$ and $v_2'$ in $V_2$ such that

$$v = v_1 + v_2 \quad \text{and} \quad v = v_1' + v_2'.$$

Then $v_1 + v_2 = v_1' + v_2'$, implying that $v_1 - v_1' = v_2 - v_2'$. But $v_1 - v_1' \, \varepsilon \, V_1$ and $v_2 - v_2' \, \varepsilon \, V_2$ since $V_1$ and $V_2$ are vector spaces. Hence $v_1 - v_1' = v_2 - v_2' \, \varepsilon \, V_1 \cap V_2$. But $V_1 \cap V_2 = \{O\}$. Therefore $v_1 - v_1' = v_2 - v_2' = O$, implying that $v_1 = v_1'$ and $v_2 = v_2'$. The expression for $v$ as the sum of elements of $V_1$ and $V_2$ is unique.

Hence $V = V_1 \oplus V_2$.

b) If $V = V_1 \oplus V_2$, it is necessary to show that $V = V_1 + V_2$ and that $V_1 \cap V_2 = \{O\}$.

By the definition of direct sum, $V = V_1 \oplus V_2 \Rightarrow V = V_1 + V_2$. Hence it remains to show that $V_1 \cap V_2 = \{O\}$.

Since both $V_1$ and $V_2$ are subspaces of $V$ each contains the zero vector and $O \, \varepsilon \, V_1 \cap V_2$.

Assume that there exists a non-zero vector $v$ in $V_1 \cap V_2$. Then,

$$v = v + O, \quad \text{where} \quad v \, \varepsilon \, V_1 \quad \text{and} \quad O \, \varepsilon \, V_2,$$

and

$$v = O + v, \quad \text{where} \quad O \, \varepsilon \, V_1 \quad \text{and} \quad v \, \varepsilon \, V_2.$$

This gives two distinct decompositions of $v$ into sums of elements of $V_1$ and $V_2$, contradicting $V = V_1 \oplus V_2$. Hence $V_1 \cap V_2 = \{O\}$.

*Theorem 4*

If $V$ is the direct sum of $V_1$ and $V_2$ then the dimension of $V$ is the sum of the dimensions of $V_1$ and $V_2$.

Proof: Let $\{v_1, \ldots, v_n\}$ and $\{v'_1, \ldots, v'_m\}$ be bases for $V_1$ and $V_2$ respectively.

If $w \varepsilon V$, there exist unique vectors $v$ and $v'$ in $V_1$ and $V_2$ respectively such that

$$w = v + v'.$$

Since scalars $\{a_i\}$ and $\{b_j\}$ exist such that $v = \sum_{i=1}^{n} a_i v_i$ and $v' = \sum_{j=1}^{m} b_j v'_j$,

$$w = \sum_{i=1}^{n} a_i v_i + \sum_{j=1}^{m} b_j v'_j. \tag{1}$$

Therefore the set $\{v_1, \ldots, v_n, v'_1, \ldots, v'_m\}$ spans $V$.

That this set is also linearly independent can be proved by contradiction. Suppose the vectors are not linearly independent. Then there exist scalars $\{c_i\}$ not all zero, such that

$$c_1 v_1 + \cdots + c_n v_n + c_{n+1} v'_1 + \cdots + c_{n+m} v'_m = O.$$

Let $c_k$ the coefficient of $v_k$ be a non-zero scalar. Then,

$$v_k = -\frac{c_1}{c_k} v_1 - \cdots - \frac{c_n}{c_k} v_n - \frac{c_{n+1}}{c_k} v'_1 - \cdots - \frac{c_{n+m}}{c_k} v'_m.$$

Consider the vector $u = \sum_{i=1}^{n} h_i v_i + \sum_{j=1}^{m} k_j v'_j$ where $h_1, \ldots, h_n, k_1, \ldots, k_n$ are non-zero scalars. Substituting for $v_k$, this vector can also be expressed

$$u = \left(h_1 - \frac{c_1}{c_k}\right) v_1 + \cdots + o v_k + \cdots + \left(h_n - \frac{c_n}{c_k}\right) v_n + \left(k_1 - \frac{c_{n+1}}{c_k}\right) v'_1$$

$$+ \cdots + \left(k_m - \frac{c_{n+m}}{c_k}\right) v'_m.$$

Hence $u$ is not uniquely expressed in terms of elements of $V_1$ and $V_2$ contradicting the assumption that $V = V_1 \oplus V_2$. Hence the set

$$\{v_1, \ldots, v_n, v'_1, \ldots, v'_m\}$$

spans $V$ and is linearly independent. It is therefore a basis for $V$ and the dimension of $V$ is $n + m$, the sum of the dimensions of $V_1$ and $V_2$.

## Exercises

1 Let $V_1$ and $V_2$ be the following sets:

$V_1 = \{(a, b, 0)$: $a$ and $b$ real numbers$\}$,
$V_2 = \{(0, c, d)$: $c$ and $d$ real numbers$\}$.

Prove that these are both subspaces of $\mathbf{R}^3$. Define $V_1 \cap V_2$ and prove that it is a subspace of $V$, illustrating theorem 1. Give the dimension and a basis for this subspace.

2 Let $V$ be a vector space and $V_1$, $V_2$ subspaces of $V$. Prove that $V_1 + V_2$ is a subspace of $V$.

3 Give an example to illustrate that the union of two subspaces need not be a subspace. We see from this question and the last why sum is a more important concept in the theory of vector spaces than is union.

4 Let $V = \mathbf{R}^2$ and let $W$ be the subspace generated by the vector (1, 1), $U$ the subspace generated by (0, 2). Prove that $V = W \oplus U$.

5 Let $V_1$ and $V_2$ be the following subsets of $\mathbf{R}^3$:

$$V_1 = \{(a, 0, b) : a, b \in \mathbf{R}\},$$

$$V_2 = \{(0, c, d) : c, d \in \mathbf{R}\}.$$

Prove that $\mathbf{R}^3 = V_1 + V_2$ but that $\mathbf{R}^3 \neq V_1 \oplus V_2$.

6 Let $U_1$ and $U_2$ be the subspaces $\{(a, a, 0) : a \in \mathbf{R}\}$ and $\{(0, b, c) : b, c \in \mathbf{R}\}$. Prove that $\mathbf{R}^3 = U_1 \oplus U_2$.

CHAPTER 2

# Matrices

## 2.1 Matrices

A *matrix* is a rectangular array with entries being elements of a field **F**. The entries are labeled thus:

$$\begin{pmatrix} a_{11} & \dots & a_{1n} \\ \vdots & & \\ a_{m1} & \dots & a_{mn} \end{pmatrix},$$

the element $a_{ij}$ being the element in the $i$th row and the $j$th column of the matrix. Such a matrix is said to be an $m \times n$ matrix over the field **F** and is denoted $(a_{ij})$. The entries will henceforth be called the *elements of the matrix*. In this chapter we develop the mathematical theory of matrices. Later we shall use matrices in a discussion of solutions to systems of linear equations and other topics.

Two matrices over the same field are said to be of the *same kind* if the number of columns in both is the same and if the number of rows in both is the same. An example of two matrices of the same kind over the field of ıeal numbers is

$$\begin{pmatrix} 1 & 2 & 3 \\ 4 & 5 & 6 \end{pmatrix} \quad \text{and} \quad \begin{pmatrix} 0 & 1 & 2 \\ 3 & 4 & 5 \end{pmatrix}.$$

These are both $2 \times 3$ matrices.

Two matrices are *equal* if they are of the same kind with corresponding elements equal.

When the number of columns $n$ in a matrix is the same as the number of rows the matrix is said to be a *square matrix of order n*. A vector with $n$ components written in row form is a $1 \times n$ matrix and written in column form is an $n \times 1$ matrix. A single element of a field **F** is a $1 \times 1$ matrix.

A matrix $B$ is the *transpose* of a matrix $A$ if the rows of $B$ are the columns of $A$. If $A$ is the matrix $(a_{ij})$ then $B$, the matrix $(b_{ij})$, is the transpose of $A$ if $a_{ij} = b_{ji}$, for all $i$ and $j$. $B$ is then denoted $A^t$.

For example if $A = \begin{pmatrix} 1 & 2 & 3 \\ 4 & 5 & 6 \end{pmatrix}$ then $A^t = \begin{pmatrix} 1 & 4 \\ 2 & 5 \\ 3 & 6 \end{pmatrix}$.

A matrix is said to be symmetric if it is equal to its transpose. Hence a symmetric matrix is necessarily square. An example of a symmetric matrix would be

$$\begin{pmatrix} 1 & 4 & 5 \\ 4 & 2 & 6 \\ 5 & 6 & 3 \end{pmatrix}.$$

The matrix $(a_{ij})$ is symmetric if $a_{ij} = a_{ji}$ for every element $a_{ij}$.

When the underlying field of scalars is the field of complex numbers a Hermitian matrix is of more significance than a symmetric matrix. A matrix is said to be *Hermitian* if it is equal to its transpose with all elements replaced by their complex conjugates. An example of such a matrix is

$$\begin{pmatrix} 1 & 1 & 3 & (4+2i) \\ 1 & 2 & -3i & 4 \\ 3 & 3i & 2 & (3-i) \\ (4-2i) & 4 & (3+i) & 2 \end{pmatrix}.$$

The significance of a Hermitian matrix will be seen in section 8.2.

The *main diagonal* of a square matrix is the set of elements $\{a_{ii}\}$. The main diagonal of the matrix $\begin{pmatrix} 1 & 2 & 3 \\ 4 & 5 & 6 \\ 7 & 8 & 0 \end{pmatrix}$ is the set $\{1, 5, 0\}$.

A *diagonal matrix* is a square matrix with non-zero elements on the main diagonal only. The matrix $\begin{pmatrix} 1 & 0 & 0 \\ 0 & 2 & 0 \\ 0 & 0 & 3 \end{pmatrix}$ is a diagonal matrix.

The operations of the field $\mathbf{F}$ enable us to define operations on the sets of matrices of the same kind over $\mathbf{F}$, under which each of these sets becomes a vector space. Let $M_{mn}$ be the set of $m \times n$ matrices over a field $\mathbf{F}$. Let $(a_{ij})$ and $(b_{ij})$ be elements of $M_{mn}$.

An operation of addition can be defined on the set $M_{mn}$ by

$$(a_{ij}) + (b_{ij}) = (a_{ij} + b_{ij}).$$

That is, addition is defined by adding corresponding elements.

As an example, $M_{23}$ is the set of $2 \times 3$ matrices over the field of real numbers, $\begin{pmatrix} 1 & 2 & 3 \\ 4 & 5 & 6 \end{pmatrix}$ and $\begin{pmatrix} 0 & 1 & 2 \\ 3 & 4 & 5 \end{pmatrix}$ are elements of $M_{23}$. The rule of addition gives

$$\begin{pmatrix} 1 & 2 & 3 \\ 4 & 5 & 6 \end{pmatrix} + \begin{pmatrix} 0 & 1 & 2 \\ 3 & 4 & 5 \end{pmatrix} = \begin{pmatrix} 1+0 & 2+1 & 3+2 \\ 4+3 & 5+4 & 6+5 \end{pmatrix} = \begin{pmatrix} 1 & 3 & 5 \\ 7 & 9 & 11 \end{pmatrix}.$$

Under this operation the set $M_{mn}$ is an Abelian group.

Next we define the operation of multiplying a matrix by a scalar, which is called *scalar multiplication of the matrix*. Let $A = (a_{ij})$ be an element of the set $M_{mn}$ and $a$ a scalar, an element of the field **F**. Then define

$$aA = (aa_{ij}).$$

That is, $aA$ is a matrix obtained by multiplying every element of $A$ by the scalar $a$.

For example, $A = \begin{pmatrix} 1 & 2 & 3 \\ 4 & 5 & 6 \end{pmatrix} \varepsilon\, M_{23}$.

Then

$$3A = \begin{pmatrix} 3 & 6 & 9 \\ 12 & 15 & 18 \end{pmatrix}.$$

Under these two operations the set $M_{mn}$ forms a vector space. Let us state this important result in the form of a theorem:

*Theorem*

Operations of addition and scalar multiplication can be defined on a set of all matrices of the same kind over a common field **F** to make this set into a vector space.

**Exercises**

1 If $A = \begin{pmatrix} 1 & 2 \\ 3 & 4 \end{pmatrix}$ and $B = \begin{pmatrix} 0 & 1 \\ 2 & 3 \end{pmatrix}$ find a) $A + B$, b) $3B$, c) $2A + 3B$.

2 Determine $A^t$ where $A$ is the matrix

a) $\begin{pmatrix} 0 & 1 \\ 2 & 3 \\ 1 & 2 \end{pmatrix}$ b) $\begin{pmatrix} 3 & 2 & 0 & 4 \\ 1 & 1 & 3 & 1 \end{pmatrix}$ c) $\begin{pmatrix} 1 & 2 & 3 \\ 4 & 5 & 6 \\ 7 & 8 & 9 \end{pmatrix}$.

3 Prove that the sum of two symmetric matrices is symmetric.

4 Prove the theorem of this section. What is the dimension of the vector space of $m \times n$ matrices? Give a basis for this space. Determine a proper subspace of this vector space.

5 Prove that the subset of $4 \times 4$ symmetric matrices forms a subspace of the vector space of all $4 \times 4$ matrices over a field **F**. What is the dimension of this subspace?

6 Give a definition of $A - B$ for two matrices $A$ and $B$ of the same kind, over the same field of scalars, that would be consistent with the definitions of this section.

7 Prove that $(A + B)^t = A^t + B^t$.

8 If $a$ is a scalar and $A$ is a matrix, prove that $(aA)^t = aA^t$.

9 Prove that $(A^t)^t = A$.

## 2.2 Multiplication of Matrices

In the previous section an operation of addition was defined between certain types of matrices, namely matrices of the same kind. In this section we define multiplication of matrices; but here again two matrices can only be multiplied together if they satisfy certain conditions.

Let $A = (a_{ij})$ be an $m \times n$ matrix and $B = (b_{ij})$ be an $n \times p$ matrix. Then define the product of $A$ and $B$, written $AB$, to be the matrix $\left(\sum_{k=1}^{n} a_{ik}b_{kj}\right)$. The product is thus a $m \times p$ matrix. Note that the number of columns in $A$ is the same as the number of rows in $B$. The product $AB$ is only defined for two such matrices.[1]

A rule that the student might find useful for evaluating the product of two matrices $A$ and $B$ is the following: to obtain the element in the $i$th row and the $j$th column of the product matrix $AB$ one takes the usual dot product of the vector that makes up the $i$th row of $A$ with the vector that makes up the $j$th column of $B$.

An example will illustrate this further. Let $A = \begin{pmatrix} 1 & 2 & 3 \\ 4 & 5 & 6 \end{pmatrix}$ and $B = \begin{pmatrix} 1 & 0 \\ 2 & 1 \\ 3 & 2 \end{pmatrix}$.

Since $A$ has three columns and $B$ has three rows the product $AB$ exists and is given by

$$\left(\begin{array}{c|c} (1\ 2\ 3)\cdot\begin{pmatrix}1\\2\\3\end{pmatrix} & (1\ 2\ 3)\cdot\begin{pmatrix}0\\1\\2\end{pmatrix} \\ \hline (4\ 5\ 6)\cdot\begin{pmatrix}1\\2\\3\end{pmatrix} & (4\ 5\ 6)\cdot\begin{pmatrix}0\\1\\2\end{pmatrix} \end{array}\right) = \left(\begin{array}{c|c} 1+4+9 & 0+2+6 \\ \hline 4+10+18 & 0+5+12 \end{array}\right)$$

$$= \begin{pmatrix} 14 & 8 \\ 32 & 17 \end{pmatrix}.$$

Note that even if $AB$ exists, $BA$ will not unless the number of columns in $B$ is equal to the number of rows in $A$. Hence the operation is not commutative.

The operations on matrices defined thus far enable us to define matrix polynomials. Consider the set of $n \times n$ matrices over a field $\mathbf{F}$. The elements of such a set can be multiplied together. Let $A$ be an arbitrary element of this set. *A polynomial in A* is an expression of the form

$$a_m A^m + a_{m-1} A^{m-1} + \cdots + a_0,$$

where $a_m, \ldots, a_0$, are scalars and $A^r = A \times \cdots \times A$, $r$ times, for any positive integer $r$. This polynomial is said to be a polynomial of degree $m$ if $a_m A^m \neq 0$. When the set of matrices is the set of $1 \times 1$ matrices over $\mathbf{R}$ then the set of polynomials is the usual set discussed in elementary algebra.

Let us now proceed to analyze further properties of matrices under the operation of multiplication.

---

[1] The student will see the motivation for this definition in question 9, exercises 3.2.

*Theorem 1*

Let $A$, $B$ and $C$ be square matrices of the same kind. Then

$$A(B + C) = AB + AC \text{ and } (A + B)C = AC + BC.$$

That is, in such a case multiplication is distributive with respect to addition.
Proof: Here $A(B + C)$ and $AB + AC$ are matrices. They will be equal if their corresponding elements are equal.

Let $A = (a_{ij})$, $B = (b_{ij})$ and $C = (c_{ij})$.

Then $B + C = (b_{ij} + c_{ij})$,

$$\begin{aligned} A(B + C) &= \Big(\sum_k a_{ik}(b_{kj} + c_{kj})\Big) \\ &= \Big(\sum_k a_{ik}b_{kj} + \sum_k a_{ik}c_{kj}\Big) \\ &= \Big(\sum_k a_{ik}b_{kj}\Big) + \Big(\sum_k a_{ik}c_{kj}\Big) \\ &= AB + AC. \end{aligned}$$

The second part of the theorem is proved similarly.

*Theorem 2*

Let $A$, $B$ and $C$ be matrices such that $AB$ and $BC$ exist. Then $(AB)C = A(BC)$.

Matrix multiplication under such conditions is associative.

Proof: Let $A = (a_{ij})$, $B = (b_{ij})$ and $C = (c_{ij})$.

Then,

$$\begin{aligned} AB &= \Big(\sum_k a_{ik}b_{kj}\Big), \\ (AB)C &= \Big(\sum_h \sum_k (a_{ik}b_{kh})\, c_{hj}\Big) \\ &= \Big(\sum_k \sum_h a_{ik}(b_{kh}c_{hj})\Big) \\ &= A(BC). \end{aligned}$$

*Theorem 3*

Let $A$ and $B$ be matrices with elements from a common field **F** and such that $AB$ exists. If $a$ is an element of **F** then $a(AB) = A(aB)$.

The student is asked to prove this result in the exercises.

It was previously seen that a set of square matrices of the same kind over a field **F** forms a vector space. Such a set also satisfies the conditions of all three of these theorems. Hence the set has, in addition to the structure of a vector space, an operation of multiplication defined on it with the properties indicated in theorems 1–3. A set with such structure is called an *Algebra*. We shall not pursue the study of this structure further in this text (students who go on to a course in functional analysis will study it there).

*Theorem 4*

Let $A$ be an $n \times n$ matrix over a field $\mathbf{F}$ which has additive identity 0 and multiplicative identity 1. Let $I_n$ be the $n \times n$ diagonal matrix

$$\begin{pmatrix} 1 & 0 & \cdots & 0 \\ 0 & 1 & \cdots & 0 \\ \vdots & & & \\ 0 & & \cdots & 0 \; 1 \end{pmatrix}.$$

Then $I_nA = AI_n = A$. $I_n$ is called the *unit* $n \times n$ matrix. It is the multiplicative identity for the set of $n \times n$ matrices over the field $\mathbf{F}$.
Proof: In the proof of this theorem we shall use the *Kronecker delta*, a very useful tool in the manipulation of matrices. It is defined thus:

$$\delta_{ab} \begin{cases} = 1 & \text{if} \quad a = b \\ = 0 & \text{if} \quad a \neq b. \end{cases}$$

Hence $I_n = (\delta_{ij})$. This gives

$$\begin{aligned} I_nA &= \left(\sum_k \delta_{ik} a_{kj}\right) \\ &= (a_{ij}) \\ &= A. \end{aligned}$$

Similarly $AI_n = A$, proving the theorem.

Let $A$ be a square matrix of order $n$. If there exists a matrix $B$ such that $AB = BA = I_n$ then $A$ is said to be invertible, $B$ is called the *inverse* of $A$ and is denoted $A^{-1}$.

*Example*

Consider the matrices

$$A = \begin{pmatrix} 3 & -1 & 1 \\ -15 & 6 & -5 \\ 5 & -2 & 2 \end{pmatrix} \quad \text{and} \quad B = \begin{pmatrix} 2 & 0 & -1 \\ 5 & 1 & 0 \\ 0 & 1 & 3 \end{pmatrix}.$$

$B$ is the inverse of $A$, since

$$\begin{pmatrix} 3 & -1 & 1 \\ -15 & 6 & -5 \\ 5 & -2 & 2 \end{pmatrix} \begin{pmatrix} 2 & 0 & -1 \\ 5 & 1 & 0 \\ 0 & 1 & 3 \end{pmatrix} = \begin{pmatrix} 2 & 0 & -1 \\ 5 & 1 & 0 \\ 0 & 1 & 3 \end{pmatrix} \begin{pmatrix} 3 & -1 & 1 \\ -15 & 6 & -5 \\ 5 & -2 & 2 \end{pmatrix} = \begin{pmatrix} 1 & 0 & 0 \\ 0 & 1 & 0 \\ 0 & 0 & 1 \end{pmatrix}$$

Later we shall give methods for determining whether or not a matrix is invertible and for finding the inverse if it exists.

If $A = (a_{ij})$ is an $n \times n$ matrix, the *trace* of $A$, denoted $tr(A)$, is the sum of the diagonal elements. That is,

$$tr(A) = a_{11} + a_{22} + \cdots + a_{nn}.$$

There are very many instances in the social sciences and business where data can best be collected and analyzed using concepts of linear algebra, as the following examples illustrate.

A doctor has four patients, each requiring three different injections at various times during the year. This data can be represented by a matrix $A$, where the element $a_{ij}$ is the number of injections of type $i$ patient $j$ is given during the year:

| | Patient 1 | Patient 2 | Patient 3 | Patient 4 |
|---|---|---|---|---|
| Injection 1 | $a_{11}$ | $a_{12}$ | $a_{13}$ | $a_{14}$ |
| Injection 2 | $a_{21}$ | $a_{22}$ | $a_{23}$ | $a_{24}$ |
| Injection 3 | $a_{31}$ | $a_{32}$ | $a_{33}$ | $a_{34}$ |

Suppose that it is known how many years each patient is to have these injections. Represent this data by a column vector $B$ with the element $b_k$ being the number of years patient $k$ is to have the injections. Then it can be seen that the elements of the column vector $AB$ represent the total numbers of each of the injections that will be given.

Further, if the cost of each injection is known, represent these costs by the elements of a row vector $C$. Then the elements of the row vector $CA$ are seen to represent the cost per year to each patient.

Social psychologists make use of matrices to record and to analyze data pertaining to group interaction. If each person in the group communicates with every other then an element $a_{ij}$ of a matrix $A$ can be used to record the reaction of person $j$ to an act from person $i$. It can be seen that the matrices used in such an analysis will necessarily be square. The diagonal elements have no interpretation and the usual convention is to label them 0.

There is a very rich relationship between the concepts of linear algebra and those of linear differential equations. For example the system

$$\begin{aligned} x_1'(t) &= a_{11}x_1(t) + \cdots + a_{1n}x_n(t) \\ &\vdots \\ x_n'(t) &= a_{n1}x_1(t) + \cdots + a_{nn}x_n(t), \end{aligned}$$

where the coefficients are functions of $t$, can be written

$$X'(t) = A(t)\,X(t),$$

where $X'(t) = \begin{pmatrix} x_1' \\ \vdots \\ x_n' \end{pmatrix}$, $X(t) = \begin{pmatrix} x_1 \\ \vdots \\ x_n \end{pmatrix}$ and $A(t) = \begin{pmatrix} a_{11} & \cdots & a_{1n} \\ \vdots & & \vdots \\ a_{n1} & & a_{nn} \end{pmatrix}$.

The set of solutions to this system forms a vector space. It is also true that the dimension of the vector space is exactly $n$. For example it can be shown that two independent solutions to the system

$$\begin{pmatrix} x_1' \\ x_2' \end{pmatrix} = \begin{pmatrix} 0 & 1 \\ 1 & 0 \end{pmatrix} \begin{pmatrix} x_1 \\ x_2 \end{pmatrix}$$

are $X_1(t) = \begin{pmatrix} e^t \\ e^t \end{pmatrix}$ and $X_2(t) = \begin{pmatrix} e^{-t} \\ -e^{-t} \end{pmatrix}$ and hence all solutions are of the form $X(t) = c_1X_1(t) + c_2X_2(t)$. The techniques that are developed in this text are used extensively in discussing existence and uniqueness of solutions and for solving systems of linear differential equations.[1]

## Exercises

1 Let $A = \begin{pmatrix} 1 & 2 \\ 3 & 4 \\ 0 & 1 \end{pmatrix}$ and $B = \begin{pmatrix} 3 & 1 & 0 & 2 \\ 4 & 2 & 1 & 3 \end{pmatrix}$. Discuss the existence of $AB$ and $BA$. Determine these products if they exist.

2 If $A = \begin{pmatrix} 0 & 1 \\ 2 & 3 \end{pmatrix}$ and $B = \begin{pmatrix} 2 & 4 \\ 0 & 1 \end{pmatrix}$ determine $A^2$, $B^2$, $AB$ and $BA$.

3 Prove that $(AB)^t = B^tA^t$.

4 Prove that if $A$, $B$ and $C$ are square matrices of the same kind then $(A + B)C = AC + BC$. This is the second part of theorem 1. Discuss the necessity of the condition that $A$, $B$ and $C$ be square matrices of the same kind for the statement of theorem 1.

5 Prove theorem 3 of this section.

6 Prove that $AI_n = A$ if $A$ is an $n \times n$ matrix. This is the second part of theorem 4.

7 If $A$ is an $m \times n$ matrix and $v$ is a column vector with $n$ components show that $Av$ exists and is a column vector with $m$ components. Further prove that if $B$ is a matrix having as columns the vectors $v_1, \ldots, v_p$, each having $n$ components, then $AB$ exists and has as columns the vectors $Av_1, \ldots, Av_p$.

8 Prove that the square of an $n \times n$ matrix is an $n \times n$ matrix. Prove further that if $A$ is an $n \times n$ matrix then $A^m$ will be an $n \times n$ matrix, where $m$ is a positive integer.

9 If $A$, $B$ and $C$ are $m \times n$, $n \times p$ and $p \times q$ matrices respectively, prove that $ABC$ is an $m \times q$ matrix.

10 Let $A = (a_{ij})$ and $B = (b_{ij})$ be two $n \times n$ matrices. Prove that

a) $tr(A + B) = tr(A) + tr(B)$.

b) $tr(AB) = \sum_{i=1}^{n} \sum_{k=1}^{n} a_{ik}b_{ki}$.

11 Let $A$ and $B$ be two $n \times n$ matrices. Prove that $tr(AB) = tr(BA)$.

12 Three companies sell four products. The number of each of the products sold by each company is known. Represent this data by a matrix $A$. If the sale price of each product is given, represent these prices by a vector and hence find the total amounts earned by each company from the sale of these products. Further, if the amount it costs each company to manufacture the products is the same for all products, represent this data by a vector and hence find the totals spent on manufacturing each product.

[1] Such systems of linear differential equations are discussed, for example, in F. Brauer and J. A. Nohel, *Qualitative Theory of Ordinary Differential Equations* (W. A. Benjamin Inc., 1969) and in D. L. Kreider, R. G. Kuller and D. R. Ostberg, *Elementary Differential Equations* (Addison-Wesley, 1969).

CHAPTER 3

# Linear Mappings

## 3.1 Mappings

In previous sections we have discussed sets and sets having the algebraic structure of a vector space. It is of interest to compare properties of sets and to relate elements of one set to another set in various ways. Conversly it is often of interest to analyse relations, such as derivatives, between sets. In this section we develop the machinery for doing this in the case of arbitrary sets and in the next section we continue in a manner relevant to vector spaces.

Let $A$ and $B$ be sets. A *mapping* from $A$ to $B$ is a rule that associates with each element of $A$ a single element of $B$. The set $A$ is called the *domain* of the mapping. We indicate that $f$ is a mapping from $A$ to $B$ by writing

$$f: A \to B.$$

If $a \varepsilon A$ and $f(a) = b$ for $b \varepsilon B$, then $b$ is the *image* of $a$ under $f$. We shall indicate this by writing $f: a \mapsto b$. If every element of $B$ is the image of at least one element of $A$ then $f$ is said to map $A$ *onto* $B$ and is called a *surjective mapping* or *onto mapping*. Otherwise, or if it is not known whether $f$ maps $A$ onto $B$ or not, we say that $f$ maps $A$ *into* $B$. The set consisting of the images of the elements of $A$ under $f$ if called the *range* of $f$.

If two distinct elements in $A$ always have distinct images in $B$ then $f$ is said to be *injective* or *one to one*.

Hence $f$ is surjective if for every $b \varepsilon B$ there exists an element $a$ of $A$ such that $f(a) = b$ and $f$ is injective if for $a_1, a_2 \varepsilon A, a_1 \neq a_2$, then $f(a_1) \neq f(a_2)$.

If $f$ is both surjective and injective it is said to be *bijective*. Such a mapping is said to have an *inverse mapping* denoted $f^{-1}$ and defined as follows: for each $b \varepsilon B$ we find the unique $a$ in $A$ such that $f(a) = b$; we then define $a$ to be $f^{-1}(b)$. This gives the value of $f^{-1}$ at each element of $B$, and thus defines $f^{-1}$ completely. $B$ is the domain of $f^{-1}$ and $A$ is its range.

The following diagrams should clarify these definitions.

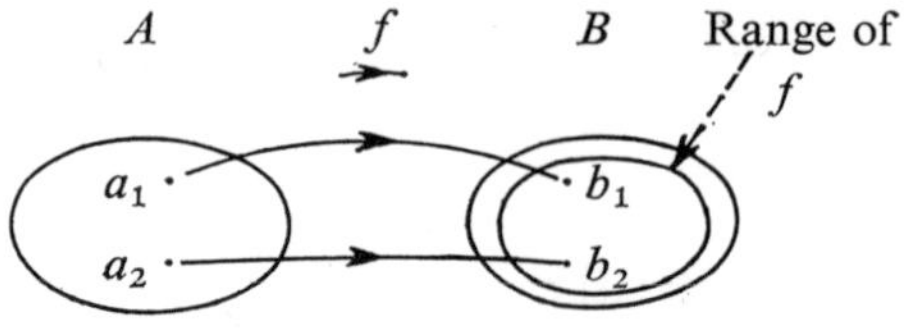


*f* is a mapping, is injective but not surjective.

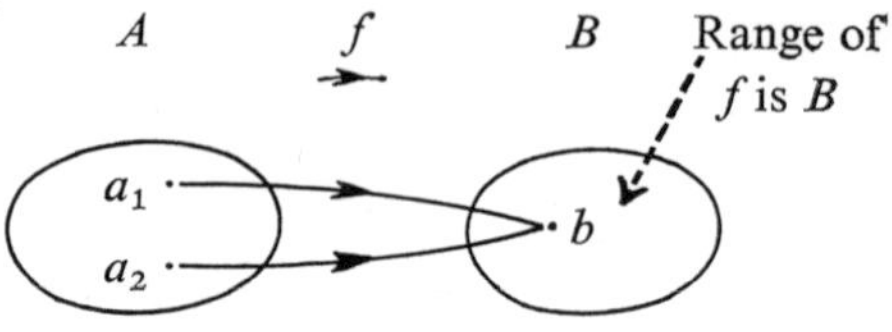


*f* is a mapping, is surjective but not injective.

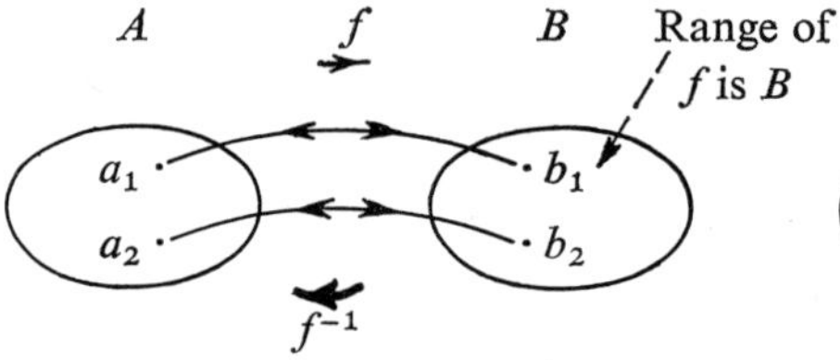


*f* is a mapping and is bijective $f^{-1}$ exists.

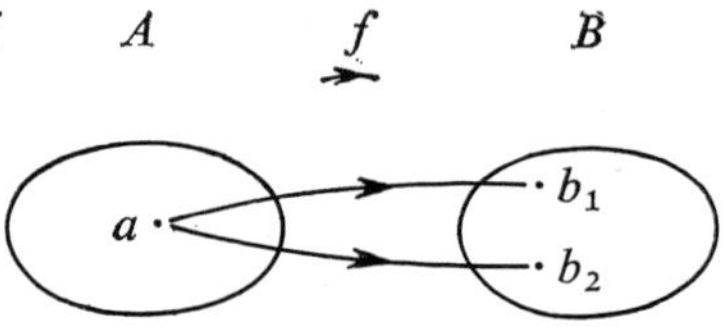


*f* is not a mapping.

*Example 1*

Let $A = \{-3, -1, 0, 1, 2\}$ and $B = \{-5, -1, 1, 3, 5\}$. Then $f$ defined by $f(a) = 2a + 1$ is a bijective mapping of $A$ onto $B$:

$f$ is a mapping since it associates with an element $a$ of $A$ the single element $2a + 1$ of $B$. For example the image of the element $-1$ of $A$ under $f$ is $-1$ in $B$, since $f(-1) = 2(-1) + 1 = -1$.

To see that $f$ is surjective consider an arbitrary element $b$ of $B$. Then we want if possible to find an element $a$ of $A$ such that $f(a) = b$, that is an element $a$ of $A$ such that $2a + 1 = b$. Such an $a$ has to satisfy the identity $a = (b - 1)/2$. It can be seen that there exists such an $a$ in $A$ for every $b \in$ of $B$; for example, when $b = 3$, then $a = (3 - 1)/2 = 1$. Hence $f[(b - 1)/2] = b$ and $f$ is surjective.

To prove that $f$ is injective consider two distinct elements $a_1$ and $a_2$ of $A$. Then $f(a_1) = 2a_1 + 1$ and $f(a_2) = 2a_2 + 1$. Since $a_1 \neq a_2$, $2a_1 + 1 \neq 2a_2 + 1$ implying that $f(a_1) \neq f(a_2)$; this proves that $f$ is injective.

$f$ is therefore bijective.

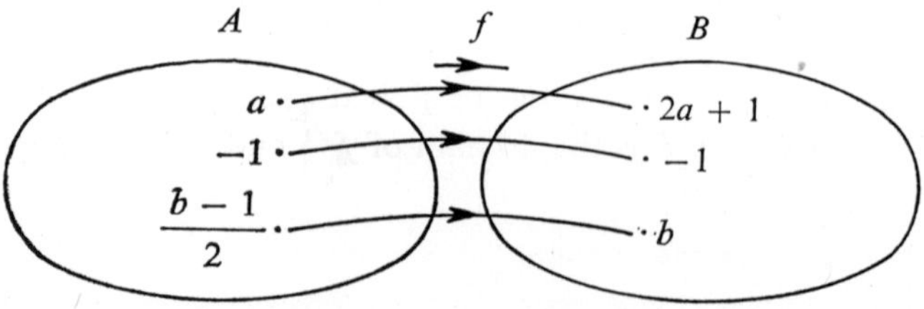

Since $f$ is bijective it has an inverse $f^{-1}$ mapping $B$ onto $A$. We have already seen that $f$ can be interpreted as the mapping that takes the element $(b - 1)/2$ of $A$ into the element $b$ of $B$. Consequently $f^{-1}$ is defined by $f^{-1}(b) = (b - 1)/2$.

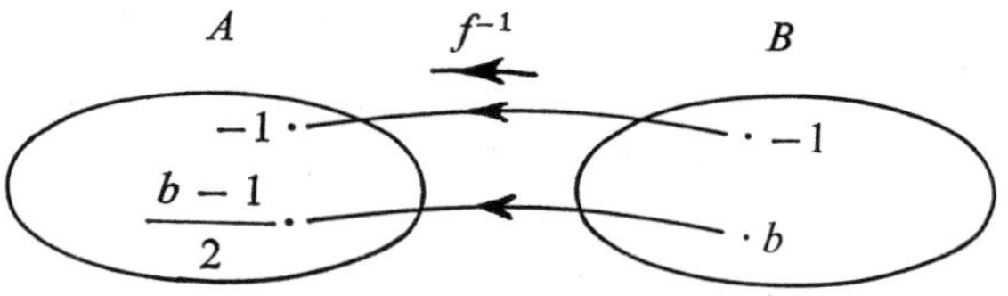


The mapping of a set onto itself that associates each element of the set with itself is called the *identity mapping*. This mapping is usually denoted by $I$ and is bijective.

A *function*[1] is a mapping of a set into the set of real numbers or the set of complex numbers. Hence the significance of functions, among mappings in general, is that the range is a subset of the real numbers or complex numbers.

*Example 2*

Let $A$ be the real line and $B$ the closed interval $[-1, 1]$ of the real line. Then $f$ defined by $f(a) = \sin a$ for every $a \varepsilon A$ is a function of $A$ into $B$ which is surjective but not injective:

$f$ is a function since with every element $a$ of $A$ there is associated a single real number in $B$, namely $\sin a$. If $b$ is an arbitrary element of $B$ then it can be considered to be the image of the set $\{a: a = \sin^{-1} b\}$ in $A$; each element of this set is mapped onto the single element $b$. Hence $f$ is surjective. It can be seen that $f$ is not injective since for example $f$ maps both 0 and $\pi$, distinct elements of $A$, onto the element 0 of $B$.

*Example 3*

Let $f$ be the mapping of $\mathbf{R}$ into itself defined by $f(x) = x^2 + 1$. Then $f$ is a function which is neither surjective nor injective:

$f$ is a function since with every element $x$ of the domain there is associated the single real number $x^2 + 1$. $f$ is not surjective since, for example, $-1$ is

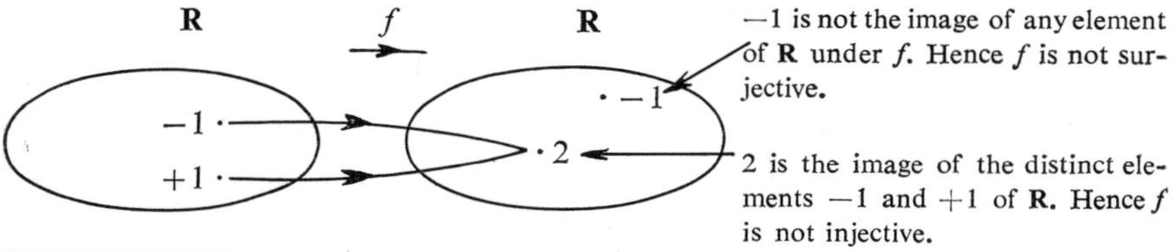


[1] Many, but not all mathematicians interpret the term *function* thus. Some mathematicians use the terms *mapping* and *function* synonymously.

not the image of any element of **R** under $f$. $f$ is not injective since, for example, the distinct elements $-1$ and $+1$ of **R** are mapped onto the single element 2. $f$ is an example of a polynomial function. A *polynomial function* is a function of the type

$$f(x) = a_n x^n + a_{n-1}x^{n-1} + \cdots + a_0, \quad (n \text{ any positive integer})$$

where the domain is the set of complex numbers or a proper subset of the complex numbers, such as **R**, the domain considered in this example. The coefficients can be real or complex numbers.

*Example 4*

The polynomial function of **R** into itself defined by $f(x) = x + 2$ is bijective:

If $y \in \mathbf{R}$, then $f(y - 2) = y$, and $y$ is the image of the element $y - 2$ under $f$. Hence $f$ is surjective. If $x_1$ and $x_2$ are distinct elements of **R**, then $x_1 + 2 \neq x_2 + 2$ implying that $f(x_1) \neq f(x_2)$; $f$ is injective. $f$ is thus bijective.

*Example 5*

Let $M$ be the set of $2 \times 2$ matrices over **R** and $f$ the mapping of $M$ into itself defined by $f(A) = A + \begin{pmatrix} 1 & 0 \\ 2 & 3 \end{pmatrix}$. $f$ is bijective:

$f$ is surjective, for if $\begin{pmatrix} a & b \\ c & d \end{pmatrix}$ is an arbitrary element of $M$, then it is the image of the matrix $\begin{pmatrix} a-1 & b \\ c-2 & d-3 \end{pmatrix}$ under $f$.

$$f\left(\begin{pmatrix} a-1 & b \\ c-2 & d-3 \end{pmatrix}\right) = \begin{pmatrix} a-1 & b \\ c-2 & d-3 \end{pmatrix} + \begin{pmatrix} 1 & 0 \\ 2 & 3 \end{pmatrix} = \begin{pmatrix} a & b \\ c & d \end{pmatrix}.$$

It is also injective for if $A$ and $B$ are distinct elements of $M$ then $A + \begin{pmatrix} 1 & 0 \\ 2 & 3 \end{pmatrix} \neq B + \begin{pmatrix} 1 & 0 \\ 2 & 3 \end{pmatrix}$, implying that $f(A) \neq f(B)$.

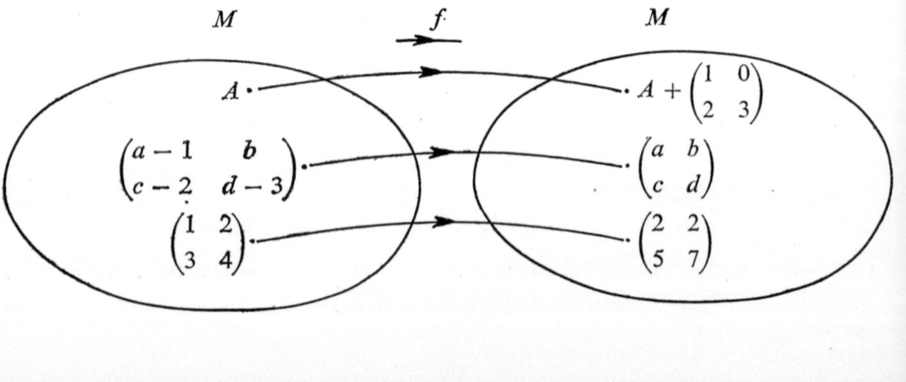

Here $f$ is an example of a *polynomial mapping* of a set of square matrices into itself. Consider the set of $n \times n$ matrices over a field **F**. Let $A$ be an arbitrary element of this set. Then $f$ defined by

$$f(A) = a_n A^n + a_{n-1} A^{n-1} + \cdots + a_0,$$

where $a_n, \ldots, a_0$ are scalars and $n$ is a positive integer, is called a polynomial mappings of the set into itself. It can be seen that for any specific $A$ in the set, $a_n A^n + a_{n-1} A^{n-1} + \cdots + a_0$, on simplification, will be an $n \times n$ matrix that is the image of $A$ under $f$.

Let $A$, $B$ and $C$ be three sets with $f$ a mapping of $A$ into $B$ and $g$ a mapping of $B$ into $C$. Then if $a \in A$, $f(a)$ will be an element of $B$ and hence will have an image $g(f(a))$ in $C$ under $g$. Thus we can associate with every element $a$ of $A$ a unique element $g(f(a))$ of $C$, using the mappings $f$ and $g$. A mapping of $A$ into $C$ is defined in this manner. We call it the *composite* of $f$ and $g$ and denote it $g \circ f$. It is the mapping

$$g \circ f(a) = g(f(a))$$

for $a \in A$.

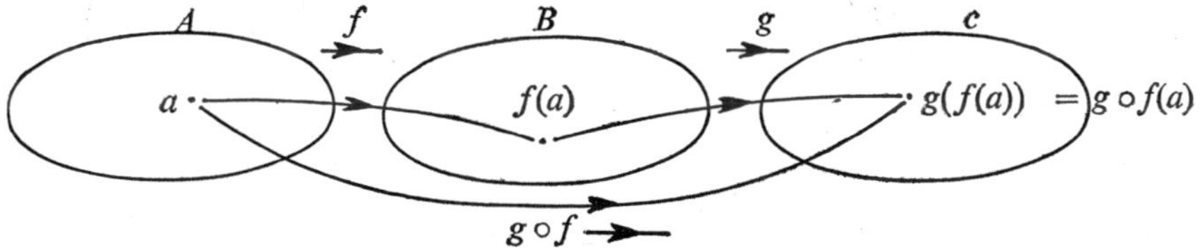


*Example 6*

Let $A = \{-1, 2, 3, 4, 5\}$, $B = \{0, 3, 4, 5, 6, 8\}$ and $C = \{0, 9, 16, 25, 36, 64, 72\}$, with $f$ defined by $f(x) = x + 1$, mapping $A$ into $B$, and $g$ defined by $g(y) = y^2$, mapping $B$ into $C$. Then $g \circ f$ is the mapping of $A$ into $C$ defined by $g \circ f(x) = (x + 1)^2$.

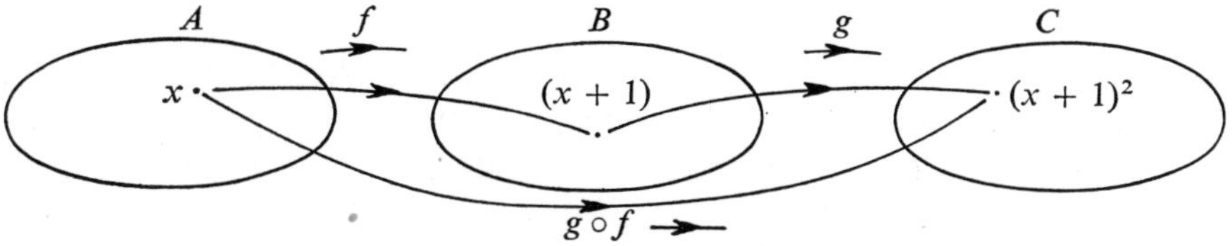


Let $f$ be a mapping of $A$ into $B$. If $C$ is a proper subset of $A$ then $f$ maps all the elements of $C$ into $B$. However, this mapping cannot be called $f$ since $f$ is a mapping with domain $A$, not $C$. This mapping of $C$ into $B$ is called the *restriction* of $f$ to $C$ and is denoted $f|_C$.

*Example 7*

Let $A = \{1, 2, 3, 4\}$, $B = \{2, 3, 4, 5, 6\}$ and $C = \{3, 4\}$ with $f$ a mapping of $A$ into $B$ defined by $f(x) = x + 2$, for $x \varepsilon A$. Then $f|_C$ is the restriction of $f$ to $C$ and is defined by $f|_C(y) = y + 2$, for $y \varepsilon C$.

Let $A$ and $B$ be sets with structures, $\cdot$ and $*$ representing operations on these two sets. A mapping $f$: $A \to B$ is said to be a *structure-preserving mapping* or a *homomorphism* with respect to these operations if the structures of $A$ and $B$ are preserved under $f$ in the following sense: if $a_1$ and $a_2$ are arbitrary elements of $A$ and $b_1$ and $b_2$ their images respectively, so that $f(a_1) = b_1$ and $f(a_2) = b_2$, then

$$f(a_1 \cdot a_2) = b_1 * b_2. \tag{1}$$

That is, if $f$ maps $a_1$ onto $b_1$ and $a_2$ onto $b_2$, it maps the *product* of $a_1$ and $a_2$ onto the *product* of $b_1$ and $b_2$ in $B$. Hence the term structure-preserving mapping.

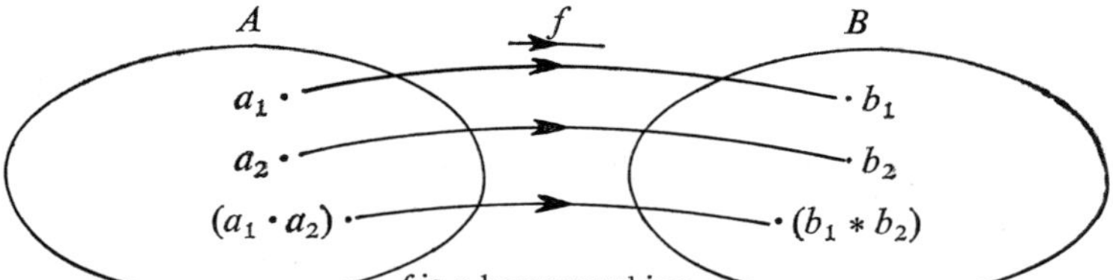


$f$ is a homomorphism

Since $b_1 = f(a_1)$ and $b_2 = f(a_2)$ it can be seen that condition (1) for a homomorphism is equivalent to

$$f(a_1 \cdot a_2) = f(a_1) * f(a_2). \tag{2}$$

If $f$ is a homomorphism which is also bijective it is said to be an *isomorphism*. In that case the spaces $A$ and $B$ are identical from the mathematical viewpoint under the correspondence defined by the isomorphism, even though the physical appearance of the elements in the two sets may differ.

*Example 8*

Let $f$ be the mapping of $\mathbf{R}^3$ into $\mathbf{R}^3$ defined by $f((a, b, c)) = (a, b, 0)$. Then $f$ preserves the operation of addition on $\mathbf{R}^3$:

If $(a, b, c)$ and $(p, q, r)$ are elements of $\mathbf{R}^3$ then $f((a, b, c)) = (a, b, 0)$ and $f((p, q, r)) = (p, q, 0)$. We have that

$$\begin{aligned} f((a, b, c) + (p, q, r)) &= f((a + p, b + q, c + r)) \\ &= (a + p, b + q, 0) = (a, b, 0) + (p, q, 0) \\ &= f((a, b, c)) + f((p, q, r)), \end{aligned}$$

demonstrating that $f$ is a structure-preserving mapping.

This mapping is an example of a *projection mapping*. It is discussed further in the next section.

*Example 9*

An isomorphism that the reader is familiar with is the logarithmic function to base 10. Let $f$ be the logarithm, mapping of the set $A$ of non-negative real numbers into the set $B$ of real numbers. Then $f$ preserves the structures of multiplication on $A$ and addition on $B$; for if $x$ and $y$ are elements of $A$, then

$$f(x) = \log_{10} x, \quad f(y) = \log_{10} y,$$

and

$$\begin{aligned} f(x \times y) &= \log_{10} (x \times y) \\ &= \log_{10} x + \log_{10} y \\ &= f(x) + f(y). \end{aligned}$$

Hence the logarithmic function is a homomorphism. Further it is a bijective mapping of $A$ onto $B$ and is therefore an isomorphism.

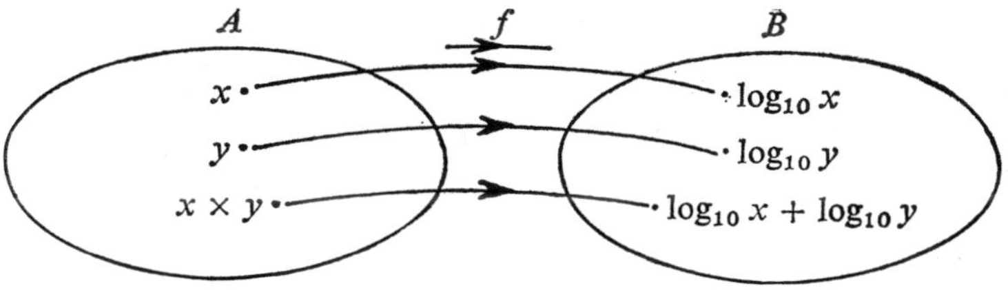


It is the fact that the logarithmic function is an isomorphism that makes it so useful. Instead of performing the operation of multiplication on the elements of the set $A$ it is easier to perform the operation of addition on the images of those elements in $B$ and then find the pre-image of the resulting element.

*Example 10*

In section 1.6 the concept of the dimension of a vector space was introduced. The dimension of a vector space is a non-negative integer that is associated with the vector space. One can think of dimension as being a function on the set of all vector spaces. Each vector space is an element of this domain space and the image of a vector space under the dimension function is the dimension of the vector space.

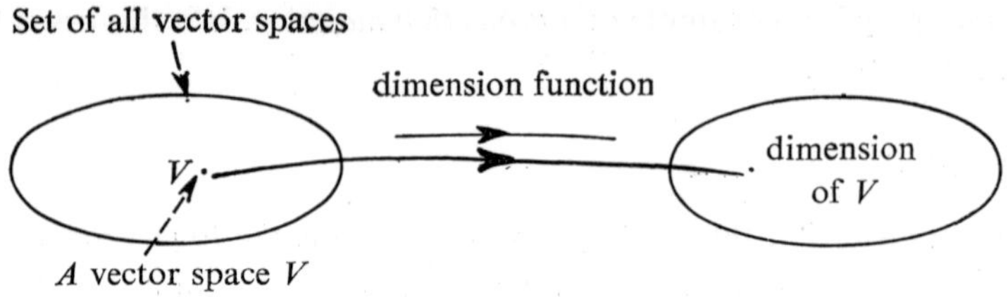


## Exercises

1 If $A$ is the interval $[0, 2\pi]$ and $B$ is the interval $[-1, 1]$ of the real line, prove that $f$ defined by $f(a) = \cos a$ is a function of $A \to B$ which is surjective but not injective.

2 Is the mapping of example 8 of this section bijective?

3 Let $f\colon \mathbf{R}^2 \to \mathbf{R}$ be the mapping defined by $f(x, y) = x + y$. Is $f$ injective? Is $f$ surjective?

4 Is the mapping $f\colon x \mapsto x^2$, of the set $\{-1, 0, 1\}$ into the set $\{0, 1\}$ injective? Surjective?

5 If $A = \{a, b, c, d, e\}$ and $B = \{p, q, r, s\}$, give examples of the following:
a) $A$ surjective mapping of $A$ onto $B$.
b) An injective mapping of a subset of $A$ into $B$.
c) A mapping of $A$ into $B$ that is neither injective nor surjective.

6 Let $P$ be the set of all polynomials defined on the interval $[0, 1]$, and having real coefficients. Define suitable operations of addition and scalar multiplication by a real number on this set that make it into a proper subspace of the space of continuous functions on this interval. Prove that the subset consisting of all such polynomials of degree less than or equal to some non-negative integer $n$ forms a further subspace. (The vector space of continuous functions on the interval $[0, 1]$ was discussed in example 2, section 1.3.)

7 Analyze the properties of the dimension function defined in example 10: What is the range; is it surjective; is it bijective?

8 If $A = \{1, 2, 3, 4, 5\}$, $B = \{0, 1, 2, 4, 9, 16, 25, 26\}$ and $C = \{-1, 0, 1, 2, 3, 8, 15, 24, 25, 28\}$, $f$ is the mapping of $A$ into $B$ defined by $f\colon x \mapsto x^2$ and $g$ is the mapping of $B$ into $C$ defined by $f\colon y \mapsto y - 1$, define $g \circ f$. What is the image of the element 4 under $g \circ f$?

9 Prove that the composite of two injective mappings is injective, whenever the composite is defined. Are similar results true for surjective mappings and homomorphisms?

10 Let $A$ and $B$ be vector spaces with $f$ a mapping of $A$ into $B$. Let $I_A$ and $I_B$ be the identity mappings of $A$ and $B$ respectively. Prove that $f$ is bijective if and only if there exists a mapping $g$ of $B$ into $A$ such that $g \circ f = I_A$ and $f \circ g = I_B$. If this is the case prove that $g = f^{-1}$.

11 We know that the set of real valued functions on the interval $[0, 1]$ can be made into a vector space $V$. Are the surjective functions from $[0, 1]$ onto $\mathbf{R}$ a subspace of $V$? Are the injective functions a subspace of $V$? Are the structure-preserving functions a subspace of $V$?

12 Prove that the mapping $f$ of $\mathbf{R}^3$ onto $\mathbf{R}^2$ defined by $f((a, b, c)) = (a, c)$ is structure-preserving in that it preserves the operations of addition on these two spaces.

13 Are the following mappings of the real line into the real line structure-preserving for the operation of addition on the real line?

a) $f(a) = a + 1$

b) $f(a) = a^2$

c) The identity mapping.

14 If $A$ is the real line and $B$ the interval $(0, \infty)$ prove that $f$ defined by $f(a) = e^a$ is a bijective mapping of $A$ onto $B$. Find its inverse. Is this mapping a homomorphism?

15 Consider the set of $n \times n$ matrices over the field of real numbers. The trace of an element of this set is a real number associated with that element. We can thus interpret *trace* as being a function on this set into **R**, the image of an element under the function being the trace of that element. Is this trace function injective? Surjective? Operations of addition are defined on both the set of $n \times n$ matrices and on **R**. Does *trace* preserve these operations?

## 3.2 Linear Mappings

In this section we shall examine mappings between vector spaces. We develop the theory of structure-preserving mappings between vector spaces. These are called *linear mappings.*

Let $U$ and $V$ be vector spaces over the same scalar field **F**. There are two operations defined on both $U$ and $V$, namely addition and scalar multiplication. A linear mapping is required to preserve these structures. Hence a mapping $f$ which maps $u$ into $f(u)$ and $v$ into $f(v)$ would have to map the sum of $u$ and $v$ into the sum of $f(u)$ and $f(v)$ in order to preserve the operation of vector addition. For $f$ to preserve the operation of scalar multiplication it would have to map the scalar multiple of $u$ by $a$, namely $au$, into the scalar multiple of $f(u)$ by $a$, namely $af(u)$. Hence a mapping $f: U \to V$ is linear if and only if

$$f(u + v) = f(u) + f(v), \text{ for every } u, v \in V. \tag{1}$$

$$f(au) = af(u) \text{ for every } u \in U \text{ and } a \in \mathbf{F}. \tag{2}$$

We illustrate these conditions by means of the following diagram.

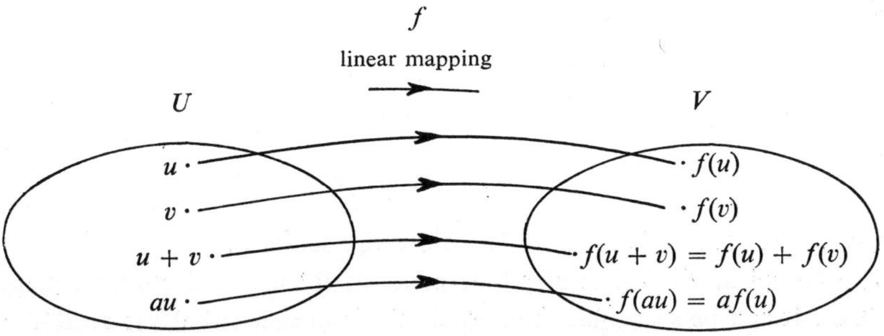

Note that linear mappings of a vector space into itself do exist. These mappings are called *operators*. The identity mapping is an operator. It is linear since

$$I(u + v) = u + v = I(u) + I(v), \text{ for every } u, v \in U,$$

$$I(au) = au = aI(u), \text{ for every } u \in U \text{ and } a \in \mathbf{F}.$$

If $U$ and $V$ are vector spaces and $f$ a linear mapping of $U$ into $V$ the set of elements in $U$ which are mapped onto the zero vector in $V$ is called the *kernel* or *null space* of the mapping.

The mapping that takes every element of $U$ into the zero element of $V$ is called the *zero mapping* of $U$ into $V$. This mapping, as we shall see in the next section, has theoretical significance.

*Example 1*

Let $f$ be the mapping that takes every element of the vector space $\mathbf{R}^3$ onto the element (0, 0) of $\mathbf{R}^2$. Since (0, 0) is the zero vector of $\mathbf{R}^2$, $f$ is the zero mapping of $\mathbf{R}^3$ into $\mathbf{R}^2$.

*Example 2*

Let $f$ be the mapping of $\mathbf{R}^3$ into $\mathbf{R}^3$ defined by

$$f: (x, y, z) \mapsto (x, y, 0).$$

$f$ is called a *projection operator* since it projects an element of $\mathbf{R}^3$ into a two-dimensional subspace of $\mathbf{R}^3$.

We now prove that $f$ is linear:

$$\begin{aligned} f((x, y, z) + (p, q, r)) &= f((x + p, y + q, z + r)) \\ &= (x + p, y + q, 0) \\ &= (x, y, 0) + (p, q, 0) \\ &= f((x, y, z)) + f((p, q, r)); \end{aligned}$$

and

$$\begin{aligned} f(a(x, y, z)) &= f((ax, ay, az)) \\ &= (ax, ay, 0) = a(x, y, 0) \\ &= af((x, y, z)), \end{aligned}$$

for every $(x, y, z)$ and $(p, q, r) \in \mathbf{R}^3$ and every scalar $a$.

The kernel of $f$ consists of all elements of $\mathbf{R}^3$ mapped onto the element (0, 0, 0). It can be seen that the kernel of $f$ is the set $\{(0, 0, z): z \in \mathbf{R}\}$.

*Example 3*

Let $V$ be the vector space of real-valued functions on $\mathbf{R}$ which have derivatives of all orders. Let $D$ be the operation of taking the derivative. Then $D: V \to V$ is linear, for

$$D(f + g) = Df + Dg,$$

and

$$D(af) = aDf,$$

for every $f$ and $g$ elements of $V$ and every real number $a$.

These properties of $D$ are known from the properties of derivatives discussed in calculus.

*Example 4*

Consider the two vector spaces $\mathbf{F}^n$ and $\mathbf{F}^m$, both over the same scalar field $\mathbf{F}$. Write the elements of $\mathbf{F}^n$ and $\mathbf{F}^m$ as column vectors. Let $A$ be an $m \times n$ matrix with elements from the field $\mathbf{F}$. Then $A$ is a linear mapping from $\mathbf{F}^n$ into $\mathbf{F}^m$ if the operation of $A$ on $\mathbf{F}^n$ is defined by the usual pre-multiplication of a vector by a matrix. That is

$$AX = \begin{pmatrix} a_{11} & \cdots & a_{1n} \\ \vdots & & \\ a_{m1} & \cdots & a_{mn} \end{pmatrix} \begin{pmatrix} x_1 \\ \vdots \\ x_n \end{pmatrix} = \begin{pmatrix} \sum_{j=1}^{n} a_{1j}x_j \\ \vdots \\ \sum_{j=1}^{n} a_{mj}x_j \end{pmatrix}.$$

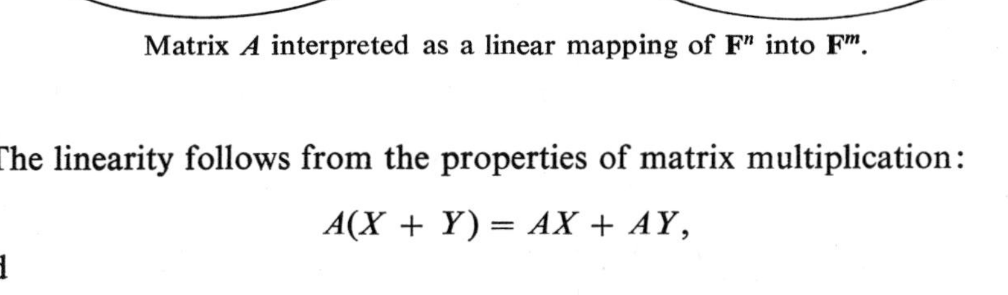


Matrix $A$ interpreted as a linear mapping of $\mathbf{F}^n$ into $\mathbf{F}^m$.

The linearity follows from the properties of matrix multiplication:

$$A(X + Y) = AX + AY,$$

and

$$A(aX) = aAX,$$

for every $X$ and $Y$ elements of $\mathbf{F}^n$ and every scalar $a$.

Hence any $m \times n$ matrix over a field $\mathbf{F}$ can be interpreted as a linear mapping of $\mathbf{F}^n$ into $\mathbf{F}^m$ with operation defined to be the operation of pre-multiplication of a vector by a matrix. We shall at times refer to a matrix as a mapping. When we do so it is to be interpreted in this manner.

We now discuss a very important class of linear mappings that is associated with every vector space of finite dimension. These mappings are called the *representation mappings* of a vector space.

Let $V$ be a vector space of dimension $n$ over a field $\mathbf{F}$ with basis $\{v_1, \dots, v_n\}$. Hence for every $v \in V$ there exist scalars $a_1, \dots, a_n$ such that $v$ can be represented

$$v = a_1 v_1 \cdots + a_n v_n.$$

Let $f$ be the mapping of $V$ into $\mathbf{F}^n$ defined by

$$f(v) = (a_1, \dots, a_n).$$

Then $f$ is linear and is called the representation mapping of $V$ associated with the basis $\{v_1, \dots, v_n\}$. *It can be seen that there is such a mapping associated with every basis of $V$.* The image of $V$ under this mapping is called the *representation* of $V$ relative to the basis $\{v_1, \dots, v_n\}$. The proof of the linearity of $f$ is as follows:

Let $v$ and $u$ be elements of $V$ with $v = \sum_{i=1}^{n} a_i v_i$ and $u = \sum_{i=1}^{n} b_i v_i$. Then $f(v) = (a_1, \dots, a_n)$ and $f(u) = (b_1, \dots, b_n)$.

We have that

$$\begin{aligned} f(v+u) &= f\left(\sum_{i=1}^{n} a_i v_i + \sum_{i=1}^{n} b_i v_i\right) \\ &= f\left(\sum_{i=1}^{n} (a_i + b_i)\, v_i\right) \\ &= (a_1 + b_1, \dots, a_n + b_n) \\ &= (a_1, \dots, a_n) + (b_1, \dots, b_n) \\ &= f(v) + f(u). \end{aligned}$$

Also

$$\begin{aligned} f(av) &= f\left(a \sum_{i=1}^{n} a_i v_i\right) \\ &= f\left(\sum_{i=1}^{n} a a_i v_i\right) = (aa_1, \dots, aa_n) \\ &= a(a_1, \dots, a_n) \\ &= af(v), \end{aligned}$$

for an arbitrary scalar $a$.

Hence $f$ is linear.
This mapping is illustrated thus:

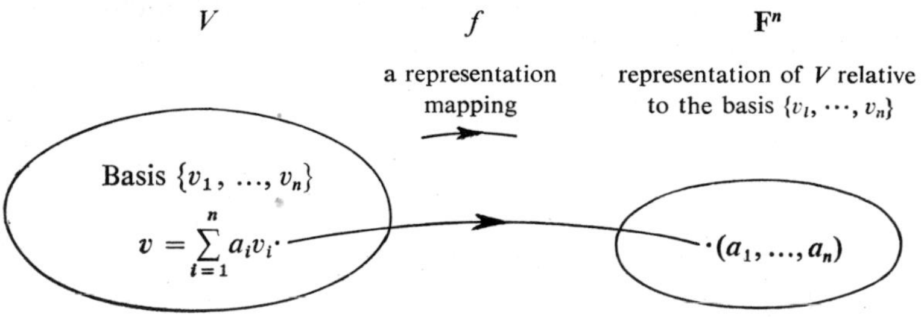


## Exercises

1 Prove that the mapping $f$: $\mathbf{R}^4 \to \mathbf{R}^4$ defined by $f(x, y, z, w) = (x, y, z, 0)$ is linear. Find the kernel of $f$. $f$ is a projection operator.

2 Prove that the zero mapping is linear.

3 Define a mapping $f$: $\mathbf{R}^3 \to \mathbf{R}^3$ by $f(x, y, z) = (x, 0, 0)$. Prove that $f$ is linear. This mapping is also a projection operator.

4 Let $f$ be the mapping of $\mathbf{R}^2$ onto $\mathbf{R}^2$ defined by $f(x, y) = (x, y) + (a, b)$, where $(a, b)$ is a fixed element of $\mathbf{R}^2$. Is $f$ linear? This is a translation on $\mathbf{R}^2$.

5 Is the following mapping linear?
$f$: $\mathbf{R}^3 \to \mathbf{R}$, defined by $f(x) = x \cdot A + 3$, where $A = (1, 0, 1)$ and $x \cdot A$ denotes the usual dot product of two vectors.

6 Prove that the mapping $A$ of $\mathbf{R}^3$ into $\mathbf{R}^3$ defined by

$$A \begin{pmatrix} x \\ y \\ z \end{pmatrix} = \begin{pmatrix} 1 & 2 & 3 \\ 0 & 1 & 2 \\ 2 & 1 & 0 \end{pmatrix} \begin{pmatrix} x \\ y \\ z \end{pmatrix}$$

is linear.

7 Let $V$ be the real vector space generated by the two functions $\sin t$, $\cos t$ on the interval $0 \leqq t \leqq \pi$, discussed in example 6, section 1.5. Let $f$ be the representation mapping of $V$ associated with the basis $\{\sin t, \cos t\}$ of $V$. Show that $f$: $V \to \mathbf{R}^2$ and that $f(\sin t) = (1, 0)$, $f(\cos t) = (0, 1)$ and $f(3 \sin t + 4 \cos t) = (3, 4)$. Find the image of $a \sin t + b \cos t$ under $f$, for arbitrary real numbers $a$ and $b$.

8 Let $V$ be the vector space of real-valued functions on $\mathbf{R}$ which have derivatives of all orders. Let $D^n$ be the operation of taking the $n^{\text{th}}$ derivative. Then $D^n$ can be interpreted as a mapping of $V$ into $V$. Prove that $D^n$ is a linear mapping.

9 Let an $m \times n$ matrix $A$ be interpreted as a linear mapping of $\mathbf{F}^n$ into $\mathbf{F}^m$ and a $p \times m$ matrix $B$ as a linear mapping of $\mathbf{F}^m$ into $\mathbf{F}^p$. Prove that $BA$ is the linear mapping of $\mathbf{F}^n$ into $\mathbf{F}^p$ that is, the composite mapping. Here we see the motivation for the definition of matrix multiplication that was given in section 2.2. Matrix multiplication was so defined that $BA$ would be the composite mapping.

## 3.3 The Space of Linear Mappings

Let $U$ and $V$ be vector spaces over a common field of scalars $\mathbf{F}$ and let $L(U; V)$ denote the set of all linear mappings from $U$ into $V$. In this section we shall define suitable operations of addition and scalar multiplication on the set $L(U; V)$ that will make it into a vector space over $\mathbf{F}$.

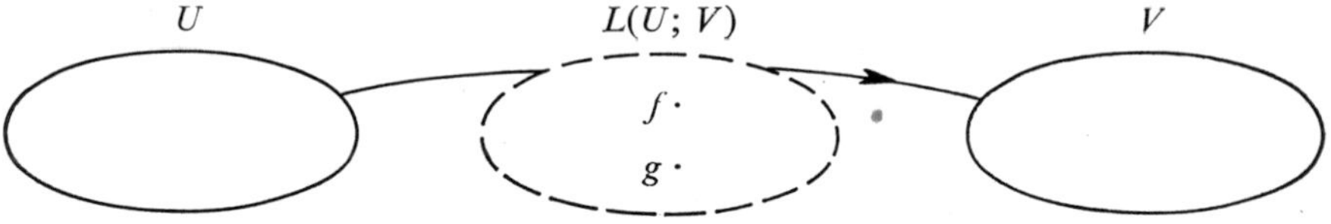


Here, and in such future diagrams, the broken curve will be used to indicate the set of mappings, so that no confusion with composite mappings will arise

First let us define an operation of addition on $L(U; V)$. If $f$ and $g$ are elements of $L(U; V)$ let $(f + g)$ be the mapping of $U$ into $V$ defined by

$$(f + g)(u) = f(u) + g(u),$$

for an arbitrary $u \varepsilon U$. The mapping $f + g$ thus defined is linear; for

$$\begin{aligned}(f + g)(u_1 + u_2) &= f(u_1 + u_2) + g(u_1 + u_2)\\ &= f(u_1) + f(u_2) + g(u_1) + g(u_2)\\ &= f(u_1) + g(u_1) + f(u_2) + g(u_2)\\ &= (f + g)(u_1) + (f + g)(u_2);\end{aligned}$$

and

$$\begin{aligned}(f + g)(au_1) &= f(au_1) + g(au_1) = af(u_1) + ag(u_1)\\ &= a[f(u_1) + g(u_1)]\\ &= a(f + g)u_1,\end{aligned}$$

for arbitrary elements $u_1$ and $u_2$ of $U$ and arbitrary scalar $a$. Hence $(f + g)$ is an element of $L(U; V)$.

This operation of addition on $L(U; V)$ is called *pointwise addition.*

The operation of scalar multiplication is defined by

$$(af)(u) = a(f(u)),$$

where $u$ is an arbitrary element of $U$ and $a$ an arbitrary scalar. It is called *pointwise scalar multiplication.* The mapping $af$ is linear.

It can be shown (left as an exercise) that all the axioms of a vector space are satisfied for $L(U; V)$ under these operations.

In future the set $L(U; V)$ will be assumed to have this vector-space structure.

## Exercises

1 Prove that the set $L(U; V)$ is a vector space under the operations defined in this section.

2 Let $V$ be the vector space of real-valued functions which have derivatives of all orders on $I$ an arbitrary interval of $\mathbf{R}$. Let $D^k$ be the operation of taking the $k^{\text{th}}$ derivative. Then $D^k$ can be interpreted as a linear mapping of $V$ into $V$ (exercise 8, section 3.2). Prove that the set of all such mappings, for $k$ a non-negative integer less than or equal to some integer $n$, generates a subspace of $L(V; V)$. This subspace of $L(V; V)$ is called the *set of constant-coefficient linear differential operators of order less than or equal to n*, on $V$. Give an example of an arbitrary element of this set.

Further, prove that the set of all elements of the form $f_n D^n + f_{n-1} D^{n-1} + \dots + f_0$, where $f_n, \dots, f_0$, are real-valued continuous functions on $I$, forms a subspace of $L(V; V)$. This subspace contains the previous one and is called the set of *linear differential operators of order less than or equal to n*, on $V$.

## 3.4 Properties of Linear Mappings

*Theorem 1*

The composite of two linear mappings is a linear mapping.

Proof: Let $U$, $V$ and $W$ be vector spaces, all over the same scalar field $\mathbf{F}$ with $f$ and $g$ linear mappings of $U$ into $V$ and $V$ into $W$ respectively. It is required to show that $g \circ f$ is a linear mapping of $U$ into $W$.

For $u_1$ and $u_2$ elements of $U$ and $a$ an element of $\mathbf{F}$, the linearity of $f$ and $g$ implies that

$$\begin{aligned} g \circ f(u_1 + u_2) &= g(f(u_1 + u_2)) = g(f(u_1) + f(u_2)) \\ &= g(f(u_1)) + g(fu_2)) \\ &= g \circ f(u_1) + g \circ f(u_2) \end{aligned}$$

and

$$\begin{aligned} g \circ f(au_1) &= g(f(au_1)) = g(af(u_1)) \\ &= ag(f(u_1)) \\ &= ag \circ f(u_1). \end{aligned}$$

Hence $g \circ f$ is a linear mapping of $U$ into $W$.

*Theorem 2*

Let $U$ and $V$ be vector spaces over the same field of scalars $\mathbf{F}$ with $\{u_1, \dots, u_n\}$ being a basis for $U$ and $f$ a linear mapping of $U$ into $V$. Then $f$ is determined completely if its value on each base vector is known. The image of $U$ under $f$ is spanned by the set $\{f(u_1), \dots, f(u_n)\}$.

Proof: Let $u \varepsilon U$, $u = \sum_{i=1}^{n} a_i u_i$. Then $f(u) = f\left(\sum_{i=1}^{n} a_i u_i\right)$. By the linearity of $f$,

$$f(u) = f\left(\sum_{i=1}^{u} a_i u_i\right) = \sum_{i=1}^{n} f(a_i u_i) = \sum_{i=1}^{n} a_i f(u_i).$$

Hence if the set $\{f(u_1), \ldots, f(u_n)\}$ is known, $f(u)$ is known and $f$ is completely determined.

Also since an arbitrary element of the image space is of the form $f(u)$ and since $f(u)$ can be expressed as a linear combination of the set

$$\{f(u_1), \ldots, f(u_n)\},$$

this set spans the image space.

We shall use this theorem later when constructing linear mappings. Defining a linear mapping on the set of base vectors will automatically define the linear mapping on the whole vector space.

*Theorem 3*

Let $U$ and $V$ be vector spaces over a common field $\mathbf{F}$ and let $f$ be a linear mapping of $U$ into $V$. Then $f(O) = O$. That is, the image of the zero vector in $U$ is the zero vector in $V$.

Note that we use $O$ to indicate the zero vector in both spaces although the two zero vectors will differ if the vector spaces are distinct. This is a notational convention and should not lead to any ambiguities if one interprets the $O$ in the correct space each time.

Proof: Let $o$ be the zero scalar. Then

$$f(O) = f(oO) = of(O) = O,$$

proving the theorem.

*Theorem 4*

Let $U$ and $V$ be vector spaces over a common field of scalars $\mathbf{F}$ and let $f$ be a linear mapping of $U$ into $V$. $f$ is injective if and only if the kernel of $f$ is the zero element of $U$.

Proof: a) Let $f$ be injective. Then for a distinct pair of elements $u_1$ and $u_2$ of $U$, $f(u_1) \neq f(u_2)$. But by the previous theorem $f(O) = O$. Hence if $u \neq O$, $f(u) \neq O$, giving that the only element that is mapped onto the zero vector of $V$ is the zero vector of $U$.

Therefore the kernel of $f$ is the zero vector of $U$.

b) Let the kernel of $f$ be $O$. Hence if $u$ is a non-zero element of $U$, $f(u) \neq O$. For $u_1$ and $u_2$ distinct elements of $U$,

$$f(u_1) - f(u_2) = f(u_1 - u_2) \neq O,$$

since $u_1 - u_2 \neq O$. Hence

$$f(u_1) \neq f(u_2).$$

Therefore $f$ is injective.

*Theorem 5*

Let $f\colon U \to V$ be injective and linear. If $u_1, \ldots, u_n$ are linearly independent vectors in $U$ then $f(u_1), \ldots, f(u_n)$ are linearly independent in $V$. That is, injective linear mappings preserve the property of linear independence.
Proof: To investigate the linear independence of the set $\{f(u_1), \ldots, f(u_n)\}$ we examine the identity

$$a_1 f(u_1) + \cdots + a_n f(u_n) = O. \tag{1}$$

This identity, by the linearity of $f$, is equivalent to

$$f(a_1 u_1 + \cdots + a_n u_n) = O,$$

which implies

$$a_1 u_1 + \cdots + a_n u_n = O$$

since $f$ is injective.

But since the set $\{u_1, \ldots, u_n\}$ is given to be linearly independent this implies that $a_1 = \cdots = a_n = o$.

Hence, returning to identity (1), the set $\{f(u_1), \ldots, f(u_n)\}$ is linearly independent.

*Converse*

Let $f$ be a linear mapping of $U$ into $V$ which preserves linear independence. Then $f$ is injective.
Proof: Let $v$ be an element of the kernel of $f$. Then $f(v) = O$. If $\{u_1, \ldots, u_n\}$ is a basis for $U$, let $v = a_1 u_1 + \cdots + a_n u_n$. Hence

$$f(a_1 u_1 + \cdots + a_n u_n) = O.$$

By the linearity of $f$ this implies that

$$a_1 f(u_1) + \cdots + a_n f(u_n) = O.$$

Since $f$ preserves linear independence the set $\{f(u_1), \ldots, f(u_n)\}$ is linearly independent, hence $a_1 = \cdots = a_n = o$. This implies that $v$ is the zero vector and hence that the mapping is injective.

The structure preserving mappings for the structure of "linear independence" are thus the injective mappings.

*Theorem 6*

Let $f$ be a linear mapping of $U$ into $V$, $W$ the range of $f$ and $K$ its kernel. Then both $W$ and $K$ are vector spaces.

Proof: Let $w_1$ and $w_2$ be elements of $W$. Then since $W$ is the range of $f$, there exist vectors $u_1$ and $u_2$ in $U$ such that

$$f(u_1) = w_1 \quad \text{and} \quad f(u_2) = w_2.$$

By the linearity of $f$ this implies that

$$w_1 + w_2 = f(u_1) + f(u_2) = f(u_1 + u_2).$$

Hence $w_1 + w_2$ is in the range of $f$ and is an element of $W$.

For any scalar $a$,

$$aw_1 = af(u_1) = f(au_1).$$

Therefore $aw_1$ is an element of $W$.

$W$ is thus a subspace of $V$ and a vector space in its own right.

It is left to the student as an exercise to prove that the kernel of a linear mapping is a vector space.

*Theorem 7*

Let $U$ be a finite dimensional vector space and $f$ a linear mapping of $U$ into a vector space $V$ with range space $W$ and kernel $K$. Then

$$\text{dimension of } U = \text{dimension of } W + \text{dimension of } K$$

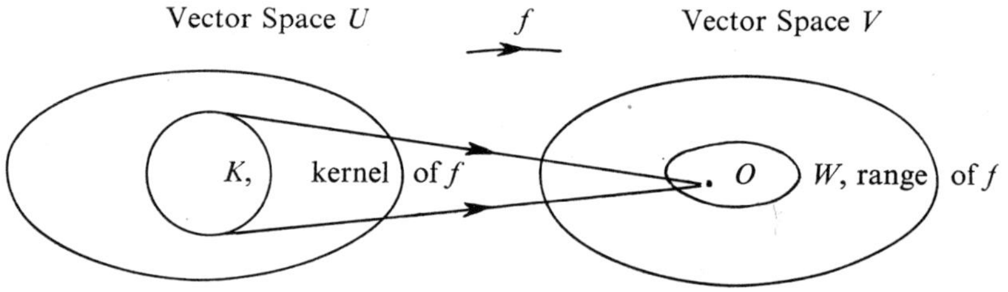


Proof: Let $\{v_1, \ldots, v_p\}$ be a basis for $K$. Complete this basis to a basis for $U$ by adjoining the vectors $v_{p+1}, \ldots, v_n$. We shall show that the vectors $f(v_{p+1}), \ldots, f(v_n)$ form a basis for $W$, thus verifying the theorem.

Let $v \in U$. Then there exist scalars $a_1, \ldots, a_n$ such that

$$v = a_1 v_1 + \cdots + a_n v_n.$$

Thus

$$f(v) = f(a_1 v_1 + \cdots + a_n v_n)$$
$$= a_1 f(v_1) + \cdots + a_p f(v_p) + a_{p+1} f(v_{p+1}) + \cdots + a_n f(v_n).$$

Since $v_1, \ldots, v_p$ are elements of the kernel of $f$, this expression reduces to

$$f(v) = a_{p+1} f(v_{p+1}) + \cdots + a_n f(v_n).$$

Hence $f(v_{p+1}), \ldots, f(v_n)$ span $W$, the range of $f$. It remains to demonstrate that these vectors are linearly independent.

Consider the identity

$$b_{p+1} f(v_{p+1}) + \cdots + b_n f(v_n) = O,$$

where $b_{p+1}, \ldots, b_n$ are arbitrary scalars, labeled thus for convenience. The linearity of $f$ implies that

$$f(b_{p+1}v_{p+1} + \cdots + b_n v_n) = O,$$

that is, that the vector $b_{p+1}v_{p+1} + \cdots + b_n v_n$ is an element of $K$. Hence it can be expressed as a linear combination of $v_1, \ldots, v_p$. Let $b_{p+1}v_{p+1} + \cdots + b_n v_n = c_1 v_1 + \cdots + c_p v_p$. But since $v_1, \ldots, v_p, v_{p+1}, \ldots, v_n$ are linearly independent (they form a basis for $U$), this equation can be satisfied only if $b_{p+1} = \cdots = b_n = c_1 = \cdots = c_p = o$. Thus $f(v_{p+1}, \ldots, f(v_n)$ are linearly independent and form a basis for $W$, proving the theorem.

A set of vectors such as $\{v_{p+1}, \ldots, v_n\}$ whose image set $\{f(v_{p+1}), \ldots, f(v_n)\}$ forms a basis for the range of $f$ is called a set of *progenitors* for the range. This definition leads to the following of theorem.

*Theorem* 8

Let $U$ be a vector space of dimension $n$ and $f$ a linear mapping of $U$ into a vector space $V$. Then $U$ is the direct sum of the kernel of $f$ and the space spanned by any set of progenitors of the range of $f$.

Proof: Let $\{u_1, \ldots, u_q\}$ be a set of progenitors of the range. Consider the identity $a_1 u_1 + \cdots + a_q u_q = O$, for arbitrary scalars $a_1, \ldots, a_q$.

This implies that

$$f(a_1 u_1 + \cdots + a_q u_q) = O,$$

and hence that

$$a_1 f(u_1) + \cdots + a_q f(u_q) = O.$$

Since $\{f(u_1), \ldots, f(u_q)\}$ is a basis for the range of $f$ the scalars $a_1, \ldots, a_q$ are identically zero implying that the set of progenitors $\{u_1, \ldots, u_q\}$, is linearly independent. Furthermore, the dimension of the range is $q$.

If $\{v_1, \ldots, v_p\}$ is a basis for the kernel of $f$ then each $u_i$, $1 \leqq i \leqq q$ is linearly independent of this set since it does not lie in the kernel. Hence $\{v_1, \ldots, v_p, u_1, \ldots, u_q\}$ is a set of $n$ linearly independent vectors in $U$. It therefore forms a basis for $U$. Any vector in $U$ can thus be written uniquely as the sum of an element of the kernel and an element of the space of progenitors, proving the theorem.

These theorems involving kernels and spaces of progenitors will be used in the discussion of the spectral theory of operators in Chapter 8.

## Exercises

1 Let $f$ and $g$ be linear mappings of $V_1$ into $V_2$ and $V_2$ into $V_3$ respectively. If $f$ and $g$ are both injective prove that the composite mapping $g \circ f$ is injective.

2 Let $f$ be the linear mapping of $\mathbf{R}^4$ into $\mathbf{R}^4$ defined by $f(a, b, c, d) = (a, d, c, 0)$. Define the kernel and the image space of this mapping and illustrate that the dimension of the kernel plus the dimension of the image space is equal to the dimension of $\mathbf{R}^4$. Find a set of progenitors for the range of this mapping and illustrate that $\mathbf{R}^4$ is the direct sum of the kernel of $f$ and the space spanned by this set.

3 Let $U$ and $V$ be finite-dimensional vector spaces and let $f$ be a linear mapping of $U$ into $V$. Denote the image of $U$ under $f$ in $V$ by $f(U)$. Prove that $\dim f(U) \leqq \dim U$. Under what condition is $\dim f(U) = \dim U$?

4 Let $f$ be a linear mapping of $U$ onto $V$. Show that any solution of the equation $f(x) = v$ for a fixed $v$ in $V$ is of the type $u + u_0$, where $u$ is an arbitrary fixed solution and $u_0$ is an element of the kernel of $f$. (This result is of importance in the discussion of solutions of linear differential equations. There $f$ is a linear differential operator, the specific solution is usually called the particular integral and the solution of $f(x) = O$ is called the complementary function. It is also of importance in a discussion of solutions to systems of linear equations, as we shall see in section 5.6.)

5 Prove that the kernel of a linear mapping is a subspace of the domain.

6 Give an example of a surjective linear mapping of $\mathbf{R}^3$ into $\mathbf{R}^2$. Prove that no injective linear mapping exists between these spaces. Generalize this result by proving that no injective linear mapping of a vector space $U$ into a vector space $V$ exists if the dimension of $U$ is greater than the dimension of $V$.

## 3.5 Isomorphisms between Vector Spaces

In section 3.1 isomorphisms between spaces with algebraic structure were defined to be structure-preserving, bijective mappings. Since in the case of vector spaces the structure-preserving mappings are the linear mappings, the isomorphisms between vector spaces are bijective linear mappings.

The vector spaces $U$ and $V$ over a common field of scalars $\mathbf{F}$ are said to be *isomorphic* if there exists a bijective linear mapping $f: U \to V$. The mapping is said to be an *isomorphism*.

A pair of isomorphic vector spaces are identical from the mathematical viewpoint when the elements of the spaces are put in correspondence using the isomorphism.

We now give a very important theorem stating the necessary and sufficient conditions for two vector spaces to be isomorphic.

*Theorem 1*

Two finite-dimensional vector spaces over the same field of scalars are isomorphic if and only if they are of the same dimension.

Proof: First let us construct an isomorphism between two vector spaces $U$ and $V$ over a field $\mathbf{F}$, having common dimension $n$.

Let $\{u_1, \ldots, u_n\}$ and $\{v_1, \ldots, v_n\}$ be bases for $U$ and $V$ respectively. If $u$ is an arbitrary element of $U$ let its representation in terms of the basis $\{u_1, \ldots, u_n\}$ be

$$u = a_1u_1 + \cdots + a_nu_n.$$

Let $f$ be a mapping of $U$ into $V$ defined by

$$f(u) = a_1v_1 + \cdots + a_nv_n.$$

This defines $f$ completely since it gives the value of $f$ on an arbitrary element of $U$. Note that $f\colon u_i \mapsto v_i$, $i = 1, \ldots, n$, for the base vectors.

We now prove that $f$ is both bijective and linear, and is thus an isomorphism. Let $w$ be an arbitrary element of $V$. Then $w$ can be expressed as a linear combination of $v_1, \ldots, v_n$,

$$w = b_1v_1 + \cdots + b_nv_n.$$

Hence $w$ is the image of the element $b_1u_1 + \cdots + b_nu_n$ of $V$ under $f$. $f$ is thus surjective.

Further, if $x$ and $y$ are distinct elements of $U$ with representations

$$x = c_1u_1 + \cdots + c_nu_n \quad \text{and} \quad y = d_1u_1 + \cdots + d_nu_n,$$

relative to the basis $\{u_1, \ldots, u_n\}$ of $V$, then

$$f(x) = c_1v_1 + \cdots + c_nv_n \quad \text{and} \quad f(y) = d_1v_1 + \cdots + d_nv_n.$$

Since $x$ and $y$ are distinct elements of $U$, $c_i \neq d_i$ for some $1 \leqq i \leqq n$. Hence $f(x) \neq f(y)$ proving that $f$ is injective.

Therefore $f$ is bijective.

It is readily seen that $f$ is linear: for vectors $x$ and $y$ of $U$,

$$\begin{aligned} f(x + y) &= f(c_1u_1 + \cdots + c_nu_n + d_1u_1 + \cdots + d_nu_n) \\ &= f((c_1 + d_1)\, u_1 + \cdots + (c_n + d_n)\, u_n) \\ &= (c_1 + d_1)\, v_1 + \cdots + (c_n + d_n)\, v_n \\ &= c_1v_1 + \cdots + c_nv_n + d_1v_1 + \cdots + d_nv_n \\ &= f(x) + f(y); \end{aligned}$$

and, for an arbitrary scalar $a$,

$$\begin{aligned} f(ax) &= f(ac_1u_1 + \cdots + ac_nu_n) \\ &= ac_1v_1 + \cdots + ac_nv_n \\ &= a(c_1v_1 + \cdots + c_nv_n) \\ &= af(x). \end{aligned}$$

Hence $f$ is an isomorphism of $U$ onto $V$.

The isomorphism is illustrated below:

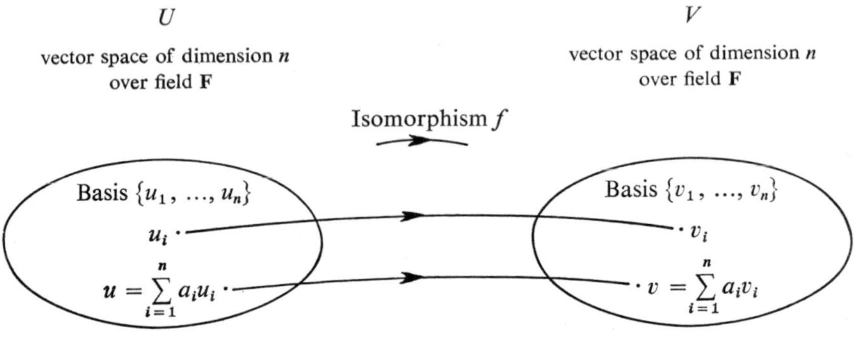


It can be seen from the construction of the above isomorphism that there exist very many isomorphisms of this type between these vector spaces. The isomorphisms depend upon the bases chosen in both spaces and further on the manner in which the base vectors are put in correspondence—this latter condition really amounts to the manner in which the base vectors are labeled.

Let us now prove the converse. Let $g$ be an isomorphism of $U$ onto $V$.

If $\{u_1, \ldots, u_n\}$ is a basis for $U$, then by theorem 5 of the previous section the set $\{g(u_1), \ldots, g(u_n)\}$ is a set of linearly independent vectors in $V$.

Further, if $v$ is an arbitrary element of $V$, since $g$ is bijective there exists an element $u$ of $U$ such that $g(u) = v$. Let

$$u = b_1u_1 + \cdots + b_nu_n.$$

Then

$$\begin{aligned} v &= g(u) \\ &= g(b_1u_1 + \cdots + b_nu_n) \\ &= b_1g(u_1) + \cdots + b_ng(u_n). \end{aligned}$$

This proves that the set $\{g(u_1), \ldots, g(u_n)\}$ also spans $V$. It is thus a basis for $V$ and the dimension of $V$ is $n$.

*Theorem 2*

The representation mappings of a vector space $U$ of dimension $n$ over a field of scalars $\mathbf{F}$ are isomorphisms of $U$ onto $\mathbf{F}^n$.

Proof: Let $\{u_1, \ldots, u_n\}$ be a basis for $U$; then it will be recalled from section 3.2 that if $u$ is an arbitrary element of $U$ and $u = a_1u_1 + \cdots + a_nu_n$, then $f$ defined by

$$f(u) = (a_1, \ldots, a_n)$$

is a representation mapping. This mapping was proved to be linear.

Let us prove that $f$ is surjective. If $(b_1, \ldots, b_n)$ is an arbitrary element of $\mathbf{F}^n$ then $b_1 u_1 + \cdots + b_n u_n$ is an element of $U$ with image $(b_1, \ldots, b_n)$ under $f$. Hence $f$ is surjective.

Next we prove that $f$ is injective. Let $u$ and $u'$ be distinct elements of $U$ with $u = \sum_{i=1}^{n} c_i u_i$ and $u' = \sum_{i=1}^{n} d_i u_i$.

Then $f(u) = (c_1, \ldots, c_n)$ and $f(u') = (d_1, \ldots, d_n)$. Since $u$ and $u'$ are distinct there exist $c_j$ and $d_j$ such that $c_j \neq d_j$. Hence $f(u) \neq f(u')$ and $f$ is injective.

$f$ is linear and bijective and is thus an isomorphism.

The spaces $U$ and $\mathbf{F}^n$ are identical mathematically when the elements are put in correspondence using $f$. When we wish to discuss $U$ or perform operations on $U$ we can do so on $\mathbf{F}^n$, which is often easier to handle than $U$, then transform the result back to $U$ under the isomorphism. Note again that the isomorphism $f$ constructed above does depend on the basis of $U$ chosen. There will exist different isomorphisms of $U$ onto $\mathbf{F}^n$ corresponding to different bases of $U$.

*Theorem 3*

The inverse of an isomorphism is an isomorphism.

Proof: Let $U$ and $V$ be isomorphic vector spaces and let $f$ be an isomorphism $U \to V$.

Since $f$ is an isomorphism it is bijective and therefore its inverse exists. Denote the inverse by $f^{-1}$. $f^{-1}$ has domain $V$ and range $U$.

Let $v_1$ and $v_2$ be distinct elements of $V$ with $f^{-1}(v_1) = u_1$ and $f^{-1}(v_2) = u_2$. This implies that $f(u_1) = v_1$ and $f(u_2) = v_2$. Since $v_1$ and $v_2$ are distinct and $f$ a mapping, $u_1 \neq u_2$. Hence $f^{-1}(v_1) \neq f^{-1}(v_2)$ and $f^{-1}$ is injective.

If $u$ is an element of $U$ then, since $f$ is a mapping of $U$ into $V$, there exists an element $v$ of $V$ such that $f(u) = v$. This implies that $f^{-1}(v) = u$. Hence $f^{-1}$ is surjective.

It remains to prove that $f^{-1}$ is linear. Let $v_1$ and $v_2$ be elements of $V$ with $f^{-1}(v_1) = u_1$ and $f^{-1}(v_2) = u_2$. This implies that $f(u_1) = v_1$ and $f(u_2) = v_2$ and that $f(u_1) + f(u_2) = v_1 + v_2$. But $f$ is linear, therefore $f(u_1 + u_2) = v_1 + v_2$, implying that $f^{-1}(v_1 + v_2) = u_1 + u_2$; that is, $f^{-1}(v_1 + v_2) = f^{-1}(v_1) + f^{-1}(v_2)$.

Further, let $f^{-1}(av) = u$ for arbitrary vector $v$ of $V$ and arbitrary scalar $a$. Then $av = f(u)$. This gives, using the linearity of $f$, $v = f(1/a\, u)$ if $a$ is non zero. Hence $f^{-1}(v) = 1/a\, u$ giving $u = af^{-1}(v)$ and $f^{-1}(av) = af^{-1}(v)$. If $a$ is the zero scalar then $av$ is zero and $f^{-1}(O) = u$. This gives $f(u) = O$. Since $f$ is linear and bijective this means that $u$ is the zero vector. Hence $f^{-1}(ov) = O$. $f^{-1}$ is linear, injective and surjective, and hence is an isomorphism.

## Exercises

1 Consider the real vector space of $2 \times 2$ matrices over the field of reals. In the exercises of section 2.1 we found that this was a real vector space of dimension 4. By the theory of this section, it is therefore isomorphic to $\mathbf{R}^4$. Define a suitable isomorphism of this vector space onto $\mathbf{R}^4$. Find the image of the element $\begin{pmatrix} 1 & 2 \\ 3 & 4 \end{pmatrix}$ under this isomorphism.

2 Let $V$ be the real vector space of continuous real-valued functions defined on the interval $0 \leqq t \leqq \pi$, $U$ the subspace of $V$ spanned by $\{\sin t, \cos t\}$, $W$ the subspace of $V$ spanned by $\{\tan t, \cos t\}$. Define an isomorphism of $U$ onto $V$.

3 Prove that the set of isomorphisms of a vector space forms a group under the operation of composition. If the vector space is of dimension $n$ this group is called the *full linear group of degree n*.

4 Let $U$ be a vector space and $f$ a linear mapping of $U$ into a vector space $V$ with range $W$. Let $P$ be a space spanned by a set of progenitors of the range of $f$. Prove that the restriction of $f$ to $P$ is an isomorphism of $P$ onto $W$.

5 Prove that the composite of two isomorphisms is an isomorphism.

6 Prove that every vector space is isomorphic to itself. Give examples of two isomorphisms of a vector space onto itself.

7 Prove that $\mathbf{R}^n$ is isomorphic to $\mathbf{R}^m$ if and only if $n = m$.

## 3.6 Bilinear and Multilinear Mappings Quadratic Forms

Let $U$, $V$ and $W$ be finite-dimensional vector spaces over a common field of scalars $\mathbf{F}$. A mapping $f$ of $U \times V$ into $W$ is said to be *bilinear* if it has the following properties:

1 $f(u_1 + u_2, v) = f(u_1, v) + f(u_2, v)$ for every $u_1, u_2 \in U$ and $v \in V$;
2 $f(u, v_1 + v_2) = f(u, v_1) + f(u, v_2)$ for every $u \in U$ and $v_1, v_2 \in V$;
3 $f(cu, v) = cf(u, v) = f(u, cv)$ for every scalar $c$ and $u \in U$, $v \in V$.

The mapping is a mapping of two variables, linear in both variables, hence the term bilinear. If $W$ is $\mathbf{F}$, the scalar field, then $f$ is usually called a *bilinear form*.

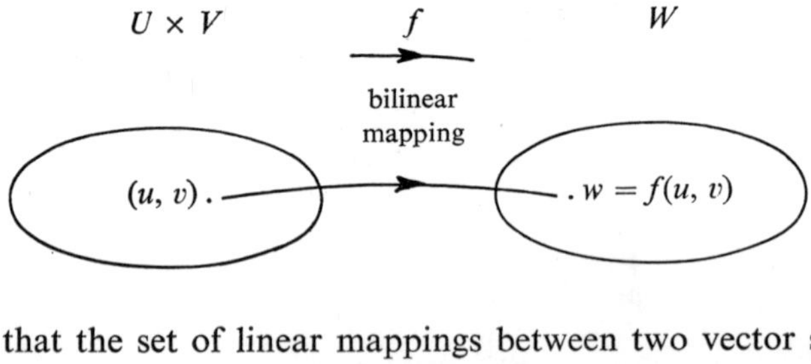


We saw that the set of linear mappings between two vector spaces over a common field of scalars could be made into a vector space by defining two suitable operations of addition and scalar multiplication. The set of all bilinear mappings of $U \times V$ into $W$ can similarly be made into a vector space.

Denote the set of bilinear mappings of $U \times V \to W$ by $B(U, V; W)$. Let $f_1$ and $f_2$ be elements of $B(U, V; W)$. Define operations of addition and scalar multiplication by elements of the scalar field $\mathbf{F}$ on the set $B(U, V; W)$ pointwise, as follows:

$$(f_1 + f_2)(u, v) = f_1(u, v) + f_2(u, v),$$

$$(cf)(u, v) = cf(u, v).$$

Under these two operations the set becomes a vector space.

If the bilinear mapping $f$ is of $U \times U$ into $W$ and has the property that $f(u_1, u_2) = f(u_2, u_1)$ for every pair of elements of $U$ it is said to be *symmetric*. If on the other hand it has the property that $f(u_1, u_2) = -f(u_2, u_1)$ it is said to be *skew-symmetric* or *anti-symmetric*.

*Example 1*

For $V$, the vector space of continuous real-valued functions over $\mathbf{R}$, let a mapping $\phi$ of $V \times V \to \mathbf{R}$ be defined by

$$\phi: (f, g) \mapsto \int_0^1 f(t)\, g(t)\, dt.$$

Then this mapping is a symmetric bilinear form:

$$\phi(f + h, g) = \int_0^1 (f(t) + h(t))\, g(t)\, dt = \int_0^1 f(t)\, g(t)\, dt + \int_0^1 h(t)\, g(t)\, dt$$

$$= \phi(f, g) + \phi(h, g),$$

$$\phi(f, k + g) = \int_0^1 f(t)\, (k(t) + g(t))\, dt = \int_0^1 f(t)\, k(t)\, dt + \int_0^1 f(t)\, g(t)\, dt$$

$$= \phi(f, k) + \phi(f, g),$$

$$\phi(cf, g) = \int_0^1 cf(t)\, g(t)\, dt = c \int_0^1 f(t)\, g(t)\, dt = c\phi(f, g).$$

Similarly $\phi(f, cg) = c\phi(f, g)$, proving that $\phi$ is a bilinear form.

Finally $\phi(f, g) = \int_0^1 f(t)\, g(t)\, dt = \int_0^1 g(t)\, f(t)\, dt = \phi(g, f)$ proving that $\phi$ is symmetric.

The concept of a bilinear mapping will now be extended in a natural way to define multilinear mappings. Let $U_i$, $i = 1, \ldots, n$ be $n$ finite-dimensional vector spaces over a common field of scalars **F** and let $V$ be another finite-dimensional vector space over **F**. An *n-linear mapping* $f$ of $U_1 \times \cdots \times U_n$ into $V$ is a mapping that has the following properties:

1 $f(u_1, \ldots, u_{i-1}, u_i + v_i, u_{i+1}, \ldots, u_n) = f(u_1, \ldots, u_{i-1}, u_i, u_{i+1}, \ldots, u_n)$
$$+ f(u_1, \ldots, u_{i-1}, v_i, u_{i+1}, \ldots, u_n),$$

for every $i = 1, \ldots, n$ and all elements of $U_1 \times \cdots \times U_n$.

2. $f(u_1 \ldots, u_{i-1}, cu_i, u_{i+1} \ldots, u_n) = cf(u_1 \ldots, u_{i-1}, u_i, u_{i+1} \ldots, u_n)$,

for every $i = 1, \ldots, n$, all elements of $U_1 \times \cdots \times U_n$ and all scalars.

The mapping is a mapping of $n$ variables which is linear in each variable. An $n$-linear mapping of $U_1 \times \cdots \times U_n$ into **F** is called an *n-linear form.*

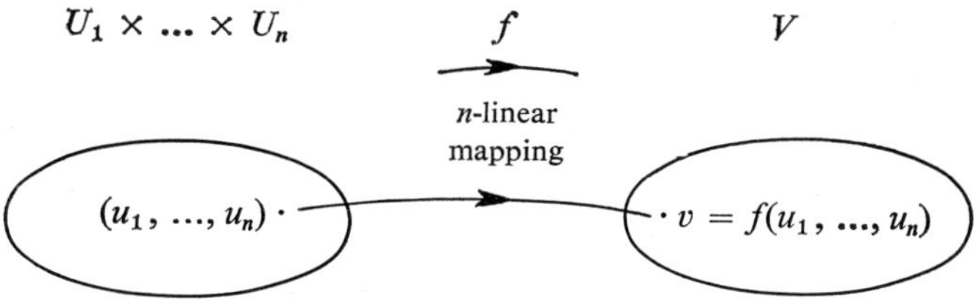


Denote the set of $n$-linear mappings of $U_1 \times \cdots \times U_n$ into $V$, $M(U_1, \ldots, U_n; V)$, $M$ indicating multilinear. Then, as in the cases of linear and bilinear mappings, this set can be made into a vector space under the following operations:

$$(f_1 + f_2)(u_1, \ldots, u_n) = f_1(u_1, \ldots, u_n) + f_2(u_1, \ldots, u_n),$$

$$(cf)(u_1, \ldots, u_n) = cf(u_1, \ldots, u_n).$$

where $f_1$ and $f_2$ are elements of $M(U_1, \ldots, U_n; V)$ and $c$ is an arbitrary scalar.

An $n$-linear mapping $f$ of $U \times \cdots \times U$ into $V$ is said to be *symmetric* if it is pairwise symmetric in the variables, that is, if

$$f(u_1, \ldots, u_i, \ldots, u_j, \ldots, u_n) = f(u_1, \ldots, u_j, \ldots, u_i, \ldots, u_n),$$

for all $i$ and $j = 1, \ldots, n$. It is *skew-symmetric* or *anti-symmetric* if

$$f(u_1, \ldots, u_i, \ldots, u_j, \ldots, u_n) = -f(u_1, \ldots, u_j, \ldots, u_i, \ldots, u_n).$$

Let $V$ be a finite-dimensional vector space over a field **F** and $f$ a symmetric bilinear form on $V$. Associated with every such bilinear form there is a mapping $q$ of $V$ into $\mathbf{F}$ called a *quadratic form*, defined thus:

$$q(v) = f(v, v).$$

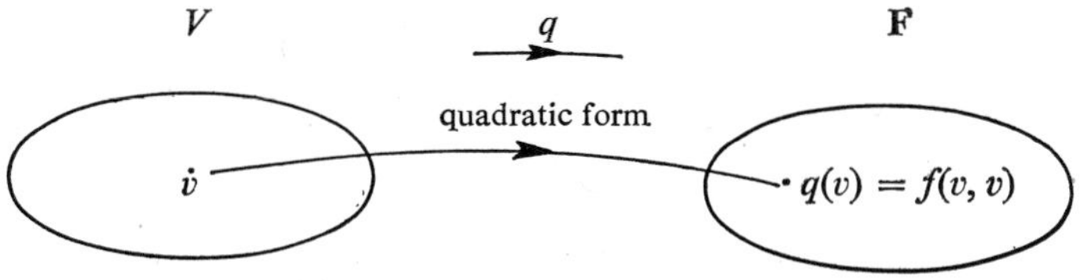


*Example 2*

Let $\phi$ be the bilinear form of example 1,

$$\phi\colon (f, g) \mapsto \int_0^1 f(t)\, g(t)\, dt.$$

It generates a quadratic form $q$ of $V$ into $\mathbf{R}$ defined by

$$q\colon f \mapsto \int_0^1 f^2(t)\, dt.$$

## Exercises

1 Prove that the set of all bilinear mappings of $U \times V$ into $W$, where $U$, $V$ and $W$ are vector spaces over a common field of scalars $\mathbf{F}$, can be made into a vector space.

2 Are the following mappings bilinear?
a) For $V$ the real vector space of continuous real-valued functions over $\mathbf{R}$, the mapping $\phi$ of $V \times V \to \mathbf{R}$ defined by $\phi(f, g) = \int_0^1 f(t)\, g(t)\, dt + 3$.
b) The mapping $\phi$ of $\mathbf{R} \times \mathbf{R} \to \mathbf{R}$ defined by $\phi(a, b) = a + b$.

3 Prove that the following mapping $f$ of $\mathbf{R} \times \mathbf{R} \times \mathbf{R} \to \mathbf{R}$ is a symmetric 3-linear mapping.

$$f(a, b, c) = \tfrac{1}{2}(abc).$$

4 Is the following mapping of $\mathbf{R} \times \mathbf{R} \times \mathbf{R} \to \mathbf{R}$ 3-linear?

$$f(a, b, c) = a^2 + b^2 + c^2.$$

5 Let $V$ be the vector space of continuous real-valued functions defined on $[0, 1]$ and let a mapping $\phi$ of $V \times V \times V \to \mathbf{R}$ be defined by

$$\phi(f, g, h) = \int_0^1 f(t)\, g(t)\, h(t)\, dt.$$

Show that this is a symmetric 3-linear mapping.

6 Let $V$ be the real vector space of continuous real-valued functions on the interval $[0, 1]$ spanned by $e^t$ and $e^{2t}$. Prove that $\phi(f, g) = \int_0^1 fg\, dt$, defines a symmetric bilinear form $\phi$ on $V$. Define the quadratic form generated by this bilinear form. Find the image of the element $2e^t + 3e^{2t}$ under this quadratic form.

7 Is it possible for two distinct symmetric bilinear forms to generate the same quadratic form? Discuss.

8 Prove that a bilinear form $f$, mapping $U \times U \to \mathbf{F}$, can be expressed as the sum of a symmetric bilinear form $g$ and a skew-symmetric bilinear form $h$, $f = g + h$.

CHAPTER 4

# Inner Product, Orthogonality, Norm

In this chapter we shall add additional structure, stage by stage, to our vector space, until we finally get a concept of magnitude for a vector.

## 4.1 Inner Product

Let $U$ be a vector space over a field $\mathbf{F}$. An *inner product*, or as it is sometimes called, a *scalar product*, is a mapping that assigns to any pair of elements $u_1$ and $u_2$ of $U$ a scalar, denoted by $\langle u_1, u_2\rangle$, satisfying the following axioms:

properties

1 $\langle u_1, u_2\rangle = \langle u_2, u_1\rangle$ for any pair of vectors $u_1$ and $u_2$ of $U$. This property is called the symmetry property of the inner product.

2 $\langle u_1, u_2 + u_3\rangle = \langle u_1, u_2\rangle + \langle u_1, u_3\rangle$ for any triple of elements $u_1$, $u_2$ and $u_3$ of $U$.

3 If $c$ is a scalar, $\langle cu_1, u_2\rangle = c\langle u_1, u_2\rangle = \langle u_1, cu_2\rangle$ for any pair of vectors $u_1$ and $u_2$ of $U$.

This scalar $\langle u_1, u_2\rangle$ is called the inner product of $u_1$ and $u_2$. There may be many mappings that satisfy these axioms. Each one defines an inner product on the space.

If in addition the inner product satisfies the condition that $\langle u, w\rangle = o$ for every $w \in U$ implies that $u = O$, the inner product is said to be *non-degenerate*. Here we shall only be interested in non-degenerate inner products. We shall assume that whenever we discuss inner products they will be non-degenerate, without having to state so.

*Example 1*

Consider the vector space $\mathbf{R}^3$ on which the usual dot product is defined. Then the dot product is an inner product for it satisfies the axioms of the inner product:
if $(a, b, c)$ and $(p, q, r)$ are elements of $\mathbf{R}^3$ then

1 $(a, b, c)\cdot(p, q, r) = ap + bq + cr = pa + qb + rc = (p, q, r)\cdot(a, b, c)$;

2 $$\begin{aligned}(a, b, c)\cdot((p, q, r) + (l, m, n)) &= (a, b, c)\cdot(p + l, q + m, r + n)\\ &= a(p + l) + b(q + m) + c(r + n)\\ &= ap + bq + cr + al + bm + cm\\ &= (a, b, c)\cdot(p, q, r) + (a, b, c)\cdot(l, m, n);\end{aligned}$$

3 For a scalar $d$,

$$d(a, b, c) \cdot (p, q, r) = (da, db, dc) \cdot (p, q, r) = dap + dbq + dcr$$
$$= d(ap + bq + cr) = d((a, b, c) \cdot (p, q, r)).$$

Similarly $(a, b, c) \cdot d(p, q, r) = d((a, b, c) \cdot (p, q, r))$.

In the standard notation, the dot product is the following inner product,

$$\langle (a, b, c), (p, q, r) \rangle = ap + bq + cr.$$

It is non-degenerate since $(a, b, c) \cdot (p, q, r) = o$ for all vectors $(p, q, r)$ implies that $ap + bq + cr = o$ for all real numbers $p, q, r$, implying that $(a, b, c) = (0, 0, 0)$ the zero vector of $\mathbf{R}^3$.

*Example 2*

Let $V$ be the real vector space of continuous real valued functions on the interval $[0, 1]$. Then an inner product is defined on $V$ by

$$\langle f, g \rangle = \int_0^1 fg\,dt.$$

It can be seen that this mapping satisfies the necessary axioms:

1 $\langle f, g \rangle = \int_0^1 fg\,dt = \int_0^1 gf\,dt = \langle g, f \rangle$ for all pairs of elements $f, g$ of $V$;

2 $\langle f, g + h \rangle = \int_0^1 f(g + h)\,dt = \int_0^1 fg\,dt + \int_0^1 fh\,dt = \langle f, g \rangle + \langle f, h \rangle$ for all triples of elements $f, g$ and $h$ of $V$;

3 $\langle af, g \rangle = \int_0^1 afg\,dt = a\int_0^1 fg\,dt = a\langle f, g \rangle$.

Similarly $\langle f, ag \rangle = a\langle f, g \rangle$, for all scalars $a$ and pairs $f$ and $g$ of $V$.

Further, $\langle f, g \rangle = o$ for all $g \in V$ implies that $\langle f, f \rangle = o$. That is $\int_0^1 f^2\,dt = o$, implying that $f = O$, the zero function and the zero element of $V$. Thus this inner product is non-degenerate.

A vector space on which an inner product has been defined will be called an *inner product space*.

We shall now relate the concept of the inner product to the bilinear form discussed in the previous section. It will be seen that every inner product on a vector space $U$ defines a symmetric bilinear form on $U \times U$ and conversely.

Let $f$ be the mapping of $U \times U$ into $\mathbf{F}$ defined by $f(u_1, u_2) = \langle u_1, u_2 \rangle$, where $\langle , \rangle$ is an inner product on $U$. We shall prove that $f$ is a symmetric bilinear form by demonstrating that it satisfies the required axioms.

1 $f(u_1 + u_2, u_3) = \langle u_1 + u_2, u_3 \rangle$

$= \langle u_3, u_1 + u_2 \rangle$, by the symmetry property of the inner product

$= \langle u_3, u_1 \rangle + \langle u_3, u_2 \rangle$, by axiom 2 for the inner product

$= \langle u_1, u_3 \rangle + \langle u_2, u_3 \rangle$

$= f(u_1, u_3) + f(u_2, u_3).$

2 $f(u_1, u_2 + u_3) = \langle u_1, u_2 + u_3 \rangle$

$= \langle u_1, u_2 \rangle + \langle u_1, u_3 \rangle$

$= f(u_1, u_2) + f(u_1, u_3).$

3 $f(cu_1, u_2) = \langle cu_1, u_2 \rangle$

$= c\langle u_1, u_2 \rangle$, by axiom 3 for the inner product

$= cf(u_1, u_2).$

$f(cu_1, u_2) = \langle cu_1, u_2 \rangle$

$= \langle u_1, cu_2 \rangle$

$= f(u_1, cu_2).$

Hence $f$ satisfies all three axioms of a bilinear form.
Further,

$$f(u_1, u_2) = \langle u_1, u_2 \rangle = \langle u_2, u_1 \rangle = f(u_2, u_1).$$

$f$ is therefore a symmetric bilinear form.

The proof of the converse, namely that every symmetric bilinear form on $U \times U$ defines an inner product on $U$, is left as an exercise for the student.

This correspondence enables us to interpret an inner product in two ways, the first as a mapping of pairs of elements of the vector space $U$ into the field of scalars and the second as a symmetric bilinear form on $U \times U$.

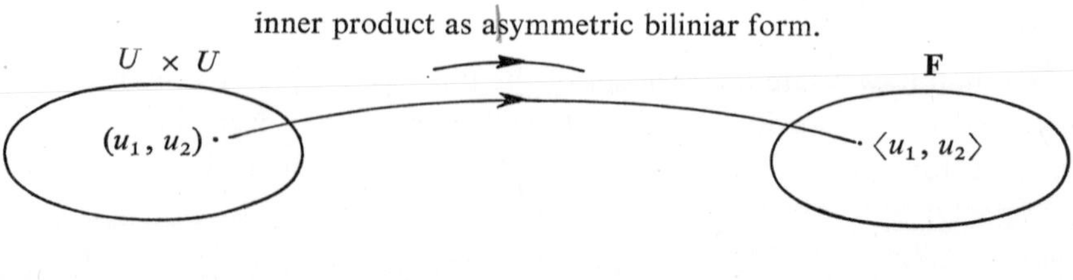

Use will be made of this second interpretation in chapters 6 and 7 to analyze further the concept of inner product.

## Exercises

1 Let $V$ be the real vector space $\mathbf{R}^2$. Prove that the usual dot product on $\mathbf{R}^2$ defines an inner product on this space.

2 Let $V$ be the real vector space having as basis the set $\{2, t\}$ of functions defined on $[0, 1]$. Prove that an inner product is defined on $V$ by

$$\langle f, g\rangle = \int_0^1 f(t)\, g(t)\, dt.$$

3 Does the mapping that associates the scalar $a + b$ with the pair of elements $a$ and $b$ of $\mathbf{R}$ define an inner product on $\mathbf{R}$?

4 Does the mapping $(f, g) \mapsto \int_0^1 f(t)\, g(t)\, dt + 3$ define an inner product on the real vector space of continuous real-valued functions?

5 Does $(a, b) \mapsto 9ab$ define an inner product on $\mathbf{R}$?

6 Does $((a, b, c), (p, q, r)) \mapsto 3ap + 3bq + 2cr$ define an inner product on $\mathbf{R}^3$?

7 If $V$ is an inner product space, $v_1, v_2, v_3$ and $v_4$ elements of $V$ and $a, b, c, d$ scalars, prove that

$$\langle av_1 + bv_2, cv_3 + dv_4\rangle = ac\langle v_1, v_3\rangle + ad\langle v_1, v_4\rangle + bc\langle v_2, v_3\rangle + bd\langle v_2, v_4\rangle,$$

using the axioms of the inner product.

8 Prove that the following is an inner product on the previously discussed real vector space of $2 \times 2$ matrices:

$$\langle A, B\rangle = \sum_{i=1}^{2} \sum_{j=1}^{2} a_{ij} b_{ij}.$$

Here $A = (a_{ij})$ and $B = (b_{ij})$.

9 Does the following mapping define an inner product on $\mathbf{R}$?

$$((a, b), (c, d)) \mapsto ad - bc.$$

10 Let $A$ be an $n \times n$ matrix. Consider the following mapping of pairs of elements of $\mathbf{R}^n$ into $\mathbf{R}$

$$((x_1, \ldots, x_n), (y_1, \ldots, y_n)) \mapsto (x_1, \ldots, x_n)\, A \begin{pmatrix} y_1 \\ \vdots \\ y_n \end{pmatrix}.$$

That is, we think of the first vector $(x_1, \ldots, x_n)$ as a $1 \times n$ matrix and the second vector $(y_1, \ldots, y_n)$ as an $n \times 1$ matrix so that the necessary matrix multiplication can be formed. Does this mapping define an inner product on $\mathbf{R}^n$? If not, what restrictions must be placed on $A$ for it to define an inner product?

11 Prove that every symmetric bilinear form on $U \times U$ defines an inner product on $U$.

12 Let $V$ be an inner product space with non-degenerate inner product and $L$ an operator on $V$. If $\langle L(u), w\rangle = o$ for $u$ and $w$ arbitrary elements of $V$ prove that $L$ is the zero operator. (The zero operator is the operator that maps every element of $V$ onto the zero vector.)

## 4.2 Orthogonality

Let $u_1$ and $u_2$ be elements of an inner product space $U$. We say that $u_1$ is *orthogonal* or *perpendicular* to $u_2$ if and only if $\langle u_1, u_2 \rangle = o$. It can be seen by the symmetry of the inner product that if $u_1$ is orthogonal to $u_2$, then $u_2$ is orthogonal to $u_1$. We denote this $u_1 \perp u_2$.

Let $U$ be an inner product space with subspaces $V$ and $W$. We can define inner products on $V$ and $W$ to be the restrictions of the inner product on $U$ to these spaces; these spaces then become inner product spaces in their own right. If for two such spaces every vector in $V$ is orthogonal to every vector in $W$ we say that *$V$ is orthogonal to $W$* and write $U \perp W$. The set of all vectors in $U$ which are orthogonal to all vectors in $V$ is called the *orthogonal complement* of $V$ and is denoted $V^\perp$.

*Example 1*

Consider the inner-product space $\mathbf{R}^3$ where the inner product is the usual dot product. Let $V$ be the two-dimensional subspace $\{(a, b, 0): a, b \in \mathbf{R}\}$ and let $W$ be the one-dimensional subspace $\{(0, 0, c): c \in \mathbf{R}\}$. Then the restriction of the inner product on $\mathbf{R}^3$ to $V$ is defined by

$$\langle (a, b, 0), (c, d, 0) \rangle = ac + bd.$$

The restriction of the inner product on $\mathbf{R}^3$ to $W$ is

$$\langle (0, 0, p), (0, 0, q) \rangle = pq.$$

It can be seen that both of these satisfy the axioms of an inner product; hence $V$ and $W$ are inner product spaces.

Let $(a, b, 0)$ and $(0, 0, c)$ be arbitrary elements of $V$ and $W$. Then under the inner product of $\mathbf{R}^3$,

$$\langle (a, b, 0), (0, 0, c) \rangle = a0 + b0 + 0c = 0.$$

Hence $V \perp W$.

Let us now find $V^\perp$. Again let $(a, b, 0)$ be an arbitrary element of $V$. Then $V^\perp = \{(l, m, n): \langle (a, b, 0), (l, m, n) \rangle = 0, l, m, n \in \mathbf{R}\}$. We have that $\langle (a, b, 0), (l, m, n) \rangle = al + bm$. This will be zero for all $a$ and $b$ elements of $\mathbf{R}$ if and only if $l = m = 0$. Hence $V = \{(0, 0, n): n \in \mathbf{R}\}$, implying that $V^\perp = W$. Similarly $W^\perp = V$.

*Example 2*

Let $V$ be the inner product space of continuous real-valued functions on the interval $[0, 1]$ with inner product defined by the mapping

$$(f, g) \mapsto \int_0^1 fg \, dt.$$

The vectors $1, \frac{1}{2} - t$ are orthogonal vectors in this space, for

$$\langle 1, \tfrac{1}{2} - t\rangle = \int_0^1 1(\tfrac{1}{2} - t)\,dt = \int_0^1 (\tfrac{1}{2} - t)\,dt$$

$$= [\tfrac{1}{2}t - t^2/2]_0^1 = 0.$$

Let $V$ be an inner product space over a field of scalars **F**, which has **R** as a subset. Then the inner product defined on $V$ is said to be *positive definite* if and only if:

$\langle v, v\rangle \geqq 0$ for every $v \varepsilon V$, and $\langle v, v\rangle = 0$ if and only if $v$ is the zero vector.

A real vector space on which a positive definite inner product has been defined is called an *Euclidean Space.*

Let $V$ be an inner product space with $\{v_1, \ldots, v_n\}$ a subset of $V$. The vectors of this subset are said to be *mutually orthogonal* if every pair of distinct vectors in the set are orthogonal, that is, if $\langle v_i, v_j\rangle = o$ for $i \neq j$. A basis consisting of a set of mutually orthogonal vectors is called an *orthogonal basis.*

*Theorem 1*

Any set of mutually orthogonal vectors of an inner product space with positive definite inner product is linearly independent.

Proof: Let $\{v_1, \ldots, v_m\}$ be such a set.

Then

$$a_1v_1 + \cdots + a_mv_m = O$$

implies that

$$\langle v_i, a_1v_1 + \ldots + a_mv_m\rangle = 0$$

for all $i = l, \ldots, m$. This identity, on account of the orthogonality of the vectors, reduces to

$$a_i\langle v_i, v_i\rangle = 0, \quad i = 1, \ldots, m.$$

Since the inner product is positive definite this implies that $a_i = 0$, $i = 1, \ldots, m$, proving that the vectors are linearly independent.

We now give a result which is of fundamental importance in the discussion of inner product spaces with positive-definite inner product.

*Theorem 2*

Let $v_1$ and $v_2$ be elements of an inner product space $V$ which has a positive definite inner product. Then

$$\langle v_1, v_2\rangle^2 \leqq \langle v_1, v_1\rangle \langle v_2, v_2\rangle.$$

This inequality is called the *Schwartz inequality.*

Proof: Let $a = -\langle v_1, v_2\rangle$ and $b = \langle v_1, v_1\rangle$. Then $av_1 + bv_2$ is an element of $V$. By the positive definiteness of the inner product,

$$\langle (av_1 + bv_2), (av_1 + bv_2)\rangle \geqq 0.$$

On expanding the inner product, usin the axioms,

$$a^2\langle v_1, v_1\rangle + b^2\langle v_2, v_2\rangle + 2ab\langle v_1, v_2\rangle \geqq 0.$$

Substituting for $a$ and $b$,

$$\langle v_1, v_2\rangle^2 \langle v_1, v_1\rangle + \langle v_1, v_1\rangle^2 \langle v_2, v_2\rangle - 2\langle v_1, v_2\rangle^2 \langle v_1, v_1\rangle = 0,$$

implying that

$$\langle v_1, v_1\rangle [\langle v_1, v_1\rangle \langle v_2, v_2\rangle - \langle v_1, v_2\rangle^2] \geqq 0.$$

If $\langle v_1, v_1\rangle = 0$, then by the positive definiteness of the inner product $v_1$ is the zero vector and the equality in the statement of the theorem holds. If $\langle v_1, v_1\rangle \neq 0$,

$$\langle v_1, v_1\rangle \langle v_2, v_2\rangle - \langle v_1, v_2\rangle^2 \geqq 0$$

and the result of the theorem still holds.

*Example 3*

Let $\mathbf{R}^2$ be the inner product space where the inner product is the usual dot product. This inner product is positive definite for if $(a, b) \in \mathbf{R}^2$ then $\langle (a, b), (a, b)\rangle = a^2 + b^2 \geqq 0$, and $\langle (a, b), (a, b)\rangle = 0$ implies that $a^2 + b^2 = 0$ and hence that $(a, b)$ is the zero vector, $(0, 0)$.

*Example 4*

Let $V$ be the inner product space of example 2. The inner product is positive definite, for $\langle f, f\rangle = \int_0^1 (f(t))^2\, dt \geqq 0$, and $\langle f, f\rangle = 0 \Rightarrow \int_0^1 (f(t))^2\, dt = 0$ $\Rightarrow f(t) = 0$, implying that $f$ is the zero function, which is the zero vector of $V$.

*Example 5*

In the inner product space $\mathbf{R}^3$ with inner product being the usual dot product the set $\{(1, 0, 0), (0, 1, 0), (0, 0, 1)\}$ forms an orthogonal basis for the space.

The student is asked to prove this result in the exercises.

## Exercises

1 In the inner product space $\mathbf{R}^2$ where the inner product is defined to be the usual dot product, prove that the two following subspaces $V$ and $W$ are orthogonal. Find $V^\perp$.

$$V = \{(x, 0): x \in \mathbf{R}\} \quad \text{and} \quad W = \{(0, y): y \in \mathbf{R}\}.$$

Interpret this result geometrically.

2 Let $\sin t$ and $\cos t$ be elements of the inner product space of continuous real-valued functions on the interval $[-\pi/2, \pi/2]$ where the inner product is defined

$$\langle f, g \rangle = \int_{-\pi/2}^{\pi/2} fg \, dt.$$

Show that these two vectors are orthogonal in this space.

3 Let $\mathbf{R}^3$ be the inner product space where the inner product is the usual dot product. Show that the vectors $(1, 0, 0)$, $(0, 1, 0)$ and $(0, 0, 1)$ are mutually orthogonal in this space and hence that this set is an orthogonal basis for the space. Prove that this inner product is positive definite.

4 Construct an alternative orthogonal basis for $\mathbf{R}^3$ to the one given in question 3.

5 Prove that $\mathbf{R}^n$ with the usual dot product is a positive-definite inner product space.

6 Let $U$ be an inner product space and $V$ a subspace of $U$. Prove that $V^\perp$ is a subpace of $U$. Note that both $V$ and $V^\perp$ are inner product spaces under the restriction of the inner product of $U$ to these sets.

7 Illustrate the Schwartz inequality by showing that in the inner product space $\mathbf{R}^3$, with inner product defined to be the dot product,

$$\langle (a, b, c), (p, q, r) \rangle^2 \leqq \langle (a, b, c), (a, b, c) \rangle \langle (p, q, r), (p, q, r) \rangle$$

for arbitrary elements $(a, b, c)$ and $(p, q, r)$ of $\mathbf{R}^3$.

8 Prove that the following mapping defines an inner product on $\mathbf{R}^4$ which is not positive definite:

$$((a, b, c, d), (p, q, r, s)) \mapsto ap + bq + cr - ds.$$

Find examples of vectors $u$ and $v$ of $\mathbf{R}^4$ that satisfy the conditions $\langle u, u \rangle = 0$ with $u \neq 0$, $\langle v, v \rangle < 0$. $\mathbf{R}^4$ with this inner product is called a *Minkowski Space*. It is the space of Special Relativity. In that theory the first three slot variables are space variables and the fourth variable represents time.

9 Find all the vectors in $\mathbf{R}^3$ that are orthogonal to $(1, 2, 3)$. Here the inner product is taken to be the dot product.

## 4.3 Norm

We now add the concept of magnitude to the elements of a vector space. Let $V$ be a vector space over the field of real numbers or complex numbers. A *norm* is a mapping that assigns to each element $v$ of $V$ a real number denoted by $\|v\|$, satisfying the following axioms.

1 $\|v\| \geqq 0$, and $\|v\| = 0$ if and only if $v$ is the zero vector.

2 $\|cv\| = |c| \, \|v\|$ for any scalar $c$. Here $|c|$ denotes the absolute value of $c$ if $c$ is a real number, modulus of $c$ if $c$ is a complex number.

3 $\|u + w\| \leqq \|u\| + \|w\|$ for any $u, w \in V$. This property is called the *triangle inequality*.

The real number $\|v\|$ is the *norm* of $v$ or the *magnitude* or *length* of $v$. Hence the term norm is used for the mapping and for the image of a vector under the mapping.

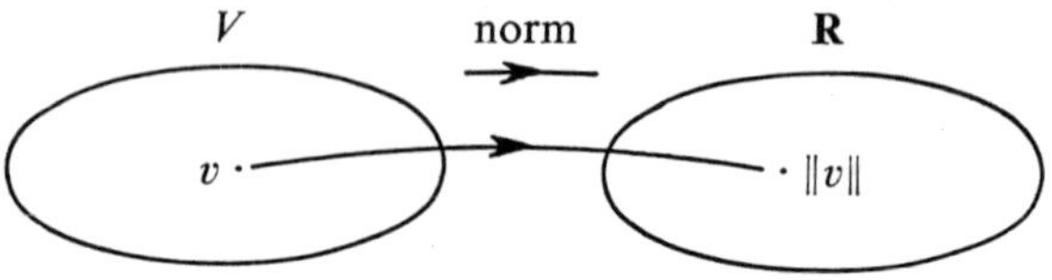


The first axiom implies that all vectors other than the zero vector have positive norm. The zero vector has zero norm.

A vector space $V$ on which a norm has been defined is said to be a *normed vector space*.

An element $v$ of $V$ is said to be a *unit vector* if and only if its norm is unity, that is, if and only if $\|v\| = 1$.

Two elements $v$ and $w$ of $V$ are said to be *parallel* if there exists a real non-zero number $c$ such that $v = cw$. The vectors are said to be in the *same direction* if the real number $c$ is greater than zero.

Let $v$ be a non-zero vector of a normed vector space $V$. Then the vector $v/\|v\|$ is a vector in the same direction as $v$ and is of unit magnitude, for $\|(v/\|v\|)\| = (1/\|v\|)\,\|v\| = 1$. This procedure of forming a unit vector in the direction of the given vector is called *normalizing* the given vector.

*Example 1*

Consider the vector space $\mathbf{R}^3$. The following mapping defines a norm on $\mathbf{R}^3$:

$$(a, b, c) \mapsto |a| + |b| + |c|.$$

To see this we check the axioms of the norm:

1 $\|(a, b, c)\| = |a| + |b| + |c| \geqq 0.$

If $(a, b, c)$ is the zero vector $a = b = c = 0$ and $\|(0, 0, 0)\| = 0$. Conversely if $\|(a, b, c)\| = 0$, $|a| + |b| + |c| = 0$, implying that $a = b = c = 0$; hence $(a, b, c)$ is the zero vector.

2 For a scalar $p$, $\|p(a, b, c)\| = \|(pa, pb, pc)\| = |pa| + |pb| + |pa|$

$= |p|\,|a| + |p|\,|b| + |p|\,|c| = |p|\,(|a| + |b| + |c|) = |p|\,\|(a, b, c)\|.$

3 $\|(a, b, c) + (p, q, r)\| = \|(a + p, b + q, c + r)\| = |a + p| + |b + q|$

$+ |c + r|.$

By the properties of real and complex numbers,

$$|a + p| \leqq |a| + |p|, |b + q| \leqq |b| + |q|, |c + r| \leqq |c| + |r|.$$

Hence $\|(a, b, c) + (p, q, r)\| \leqq |a| + |p| + |b| + |q| + |c| + |r| = \|(a, b, c)\| + \|(p, q, r)\|$.
Therefore all the axioms of a norm are satisfied.

If $V$ is an inner product space with a positive definite inner product $\langle , \rangle$ there exists a norm of particular interest on this space. There may of course exist many norms on this space, but there is one norm generated by the inner product. This is called the *usual norm* or the *norm induced by the inner product*, and is defined thus for $v \in V$:

$$\|v\| = \sqrt{\langle v, v \rangle}.$$

It is necessary to demonstrate that this definition satisfies the axioms of a norm:

1 $\|v\| = \sqrt{\langle v, v \rangle} \geqq 0.$

If $v$ is the zero vector, $\langle v, v \rangle = 0$ by the positive definiteness of the inner product, implying that $\|v\| = 0$. Conversely if $\|v\| = 0$, $\langle v, v \rangle = 0$ implying that $v$ is the zero vector.

2 $\|cv\| = \sqrt{\langle cv, cv \rangle} = \sqrt{c^2 \langle v, v \rangle} = |c| \sqrt{\langle v, v \rangle} = |c| \, \|v\|.$

3 $\|v + w\| = \sqrt{\langle v + w, v + w \rangle} = \sqrt{\langle v, v \rangle + \langle v, w \rangle + \langle w, v \rangle + \langle w, w \rangle}.$

Hence $\|v + w\|^2 = \langle v, v \rangle + \langle v, w \rangle + \langle w, v \rangle + \langle w, w \rangle = \|v\|^2 + 2\|v, w\| + \|w\|^2$.
We now apply the Schwartz inequality to get

$$\|v + w\|^2 \leqq \|v\|^2 + 2\langle v, v \rangle^{1/2} \langle w, w \rangle^{1/2} + \|w\|^2.$$

This may be written

$$\|v + w\|^2 \leqq \|v\|^2 + 2\|v\| \, \|w\| + \|w\|^2,$$

giving

$$\|v + w\|^2 \leqq (\|v\| + \|w\|)^2,$$

proving the triangle inequality property.

*Example 2*

We know from section 4.1 that $\mathbf{R}^3$ with the usual dot product is an inner product space. Here we find the norm induced by this inner product.

If $(a, b, c) \in \mathbf{R}^3$ then

$$\langle (a, b, c), (a, b, c) \rangle = a^2 + b^2 + c^2.$$

Hence if $\| \; \|$ is the usual norm for $\mathbf{R}^3$, it is defined by

$$\|(a, b, c)\| = \sqrt{\langle (a, b, c), (a, b, c) \rangle} = \sqrt{a^2 + b^2 + c^2}.$$

This is the usual concept for the magnitude of a vector in $\mathbf{R}^3$.

If $V$ is a normed inner product space with the usual norm, a basis $\{v_1, \ldots, v_n\}$ of $V$ is said to be *orthonormal* if it is an orthogonal basis and if each base vector is a unit vector.

*Example 3*

Consider the normed inner product space $\mathbf{R}^3$ with the usual norm, as defined in example 2. Then the set $\{(1, 0, 0), (0, 1, 0), (0, 0, 1)\}$ is an orthonormal basis for this space. The proof of this example is left as an exercise.

## Exercises

1 Find an orthonormal basis for the normed inner product space $\mathbf{R}^2$, where the inner product is the dot product and the norm is the induced norm.

2 The mapping $((a, b, c), (p, q, r)) \mapsto 3ap + 3bq + 2cr$ defines an inner product on $\mathbf{R}^3$ [exercise 6, section 4.1]. Prove that this inner product is positive definite and determine the magnitudes of the vectors $(1, 0, 0)$, $\left(\frac{1}{\sqrt{3}}, 0, 0\right)$ and $(1, 2, 3)$ under the norm induced by this inner product.

3 Consider the real vector space of $2 \times 2$ matrices with inner product defined by $\langle A, B\rangle = \sum_{i=1}^{2} \sum_{j=1}^{2} a_{ij}b_{ij}$ [exercise 8, section 4.1]. Prove that this inner product is positive definite. Find an orthonormal basis for this space, the norm being the norm induced by this inner product.

4 Prove that the following function defines a norm on the real vector space of $2 \times 2$ matrices:

$$f\colon A \mapsto \sum_{i=1}^{2} \sum_{j=1}^{2} |a_{ij}|,$$

where $A = (a_{ij})$.[1]

5 Let $V$ be the inner product space of continuous real-valued functions on the interval $[0, 1]$ with inner product defined by $\langle f, g\rangle = \int_0^1 fg\, dt$ [see example 2, section 4.1]. Find the induced norm on this space and hence determine the magnitudes of the vectors $t^2, t^4, t + 1$.

6 Prove that two inner products defined on a vector space are identical if and only if the norms that they induce are the same [You may use exercise 7, section 3.6].

## 4.4 The Gram-Schmidt Orthogonalization Process

Let $V$ be a finite-dimensional normed inner product space with positive definite inner product. The norm is taken to be the usual norm. In this section we prove that there always exists an orthonormal basis for such a space and show how it can be constructed.

---

[1] Such matrix norms are used in deriving criteria for the existence and uniquenesd of solutions of linear systems of differential equations. See, for example, F. Brauer ans J. A. Nohel, *Qualitative Theory of Ordinary Differential Equations*, section 2.2 (W. A. Benjamin Inc., 1969).

Let $\{v_1, \ldots, v_n\}$ be an arbitrary basis for $V$. First of all we construct an orthogonal basis from this space and then obtain an orthonormal basis by normalizing each vector of the orthogonal basis. Denote the orthogonal basis $\{u_1, \ldots, u_n\}$.

There is complete freedom in the choice of the first vector of the orthogonal basis; let $u_1 = v_1$.

Next consider the vector

$$v_2 - \frac{\langle v_2, u_1\rangle}{\langle u_1, u_1\rangle} u_1 .$$

Using the properties of the inner product, since $\langle u_1, u_1\rangle = \|u_1\|^2$, the vector can be written

$$v_2 - \left\langle v_2, \frac{u_1}{\|u_1\|}\right\rangle \frac{u_1}{\|u_1\|} .$$

Here $u_1/\|u_1\|$ is a unit vector in the direction of $u_1$ and $\langle v_2, u_1/\|u_1\|\rangle$ is the inner product of $v_2$ with this unit vector. The vector $\langle v_2, u_1/\|u_1\|\rangle\, u_1 \|u_1\|$ is quite naturally called the *projection of* $v_2$ *in the direction of* $u_1$. Its magnitude is $\langle v_2, u_1/\|u_1\|\rangle$. The projection of $v_2$ in the direction of $u_1$ is in effect the component of $v_2$ in the direction of $u_1$. If $v_2$ is orthogonal to $u_1$ its projection in the direction of $u_1$ is seen to be zero.

We now prove that $v_2 - \langle v_2, u_1/\|u_1\|\rangle\, u_1/\|u_1\|$ is orthogonal to $u_1$. That is, as one would expect, on subtracting the projection of $v_2$ in the direction of $u_1$ away from $v_2$, the remainder is orthogonal to $u_1$. We have that

$$\left\langle v_2 - \left\langle v_2, \frac{u_1}{\|u_1\|}\right\rangle \frac{u_1}{\|u_1\|}, u_1 \right\rangle = \langle v_2, u_1\rangle - \left\langle v_2, \frac{u_1}{\|u_1\|}\right\rangle \left\langle \frac{u_1}{\|u_1\|}, u_1\right\rangle$$

$$= \langle v_2, u_1\rangle - \langle v_2, u_1\rangle \frac{\langle u_1, u_1\rangle}{\|u_1\|^2} = 0,$$

proving the result.

Hence let

$$u_2 = v_2 - \frac{\langle v_2, u_1\rangle}{\langle u_1, u_1\rangle} u_1 .$$

We now have two members of the orthogonal basis. Continuing the process, it can be shown that $u_3$ defined by

$$u_3 = v_3 - \frac{\langle v_3, u_1\rangle}{\langle u_1, u_1\rangle} u_1 - \frac{\langle v_3, u_2\rangle}{\langle u_2, u_2\rangle} u_2$$

is perpendicular to both $u_1$ and $u_2$. Here the projections of $v_3$ in the directions of both $u_2$ and $u_1$ have been subtracted away from it. The resulting vector $u_3$ is perpendicular to both $u_1$ and $u_2$. This process is continued.

In general the $i$th vector of the orthogonal basis is defined inductively in terms $u_1, \ldots, u_{i-1}$ and is

$$u_i = v_i - \frac{\langle v_i, u_1 \rangle}{\langle u_1, u_1 \rangle} u_1 - \cdots - \frac{\langle v_i, u_{i-1} \rangle}{\langle u_{i-1}, u_{i-1} \rangle} u_{i-1}.$$

The set $\{u_1, \ldots, u_n\}$ thus formed is orthogonal. Normalizing each of these vectors now gives an orthonormal basis.

*Example 1*

Consider the normed inner product space $\mathbf{R}^3$ with inner product defined by the dot product and norm being the induced norm. The set

$$\{(3, 0, 0), (0, 1, 2), (0, 2, 5)\}$$

is a basis for this space. Using the Schmidt orthogonalization process construct an orthonormal basis from this basis.

Using the notation of the above theory, let $v_1 = (3, 0, 0)$, $v_2 = (0, 1, 2)$ and $v_3 = (0, 2, 5)$.

Then $u_1 = v_1 = (3, 0, 0)$.

$$u_2 = v_2 - \frac{\langle v_2, u_1 \rangle u_1}{\langle u_1, u_1 \rangle} = (0, 1, 2) - \left[\frac{(0, 1, 2) \cdot (3, 0, 0)}{(3, 0, 0) \cdot (3, 0, 0)}\right](3, 0, 0)$$
$$= (0, 1, 2).$$

Here $v_2$ is orthogonal to $u_1$ hence its projection in the direction of $u_1$ is zero and $u_2$ can be taken to be $v_2$.

$$u_3 = v_3 - \frac{\langle v_3, u_1 \rangle u_1}{\langle u_1, u_1 \rangle} - \frac{\langle v_3, u_2 \rangle u_2}{\langle u_2, u_2 \rangle}$$
$$= (0, 2, 5) - \left[\frac{(0, 2, 5) \cdot (3, 0, 0)}{(3, 0, 0) \cdot (3, 0, 0)}\right](3, 0, 0) - \left[\frac{(0, 2, 5) \cdot (0, 1, 2)}{(0, 1, 2) \cdot (0, 1, 2)}\right](0, 1, 2)$$
$$= (0, -2/5, 1/5).$$

The set $\{(3, 0, 0), (0, 1, 2), (0, -2/5, 1/5)\}$ forms an orthogonal basis for $\mathbf{R}^3$.

On normalizing each of these vectors we have that the set

$$\{(1, 0, 0), (0, 1/\sqrt{5}, 2/\sqrt{5}), (0, -2/\sqrt{5}, 1/\sqrt{5})\}$$

forms an orthonormal basis for the space.

*Theorem 1*

Let $V$ be a normed inner product space over the field of real numbers having a positive definite inner product. If $W$ is a subspace of $V$,

$$\dim V = \dim W + \dim W^{\perp}.$$

Proof: If $W = V$ then $W^\perp = \{O\}$ and $\dim V = \dim W$, $\dim W^\perp = 0$, and the result is true. Similarly the result is true if $W = \{O\}$.

It remains to consider the case of $W$ being a proper subspace of $V$ and $\neq \{O\}$.

Let $\{w_1, \ldots, w_r\}$ be an orthonormal basis for $W$. This can be extended to give a basis $\{w_1, \ldots, w_r, v_1, \ldots, v_m\}$ for $V$. Using the Schmidt orthogonalization process this basis can be used to generate an orthonormal basis $\{w_1, \ldots, w_r, u_1, \ldots, u_m\}$ for $V$. Note that since the vectors $w_1, \ldots, w_r$ are given to be orthonormal the Schmidt orthogonalization process leaves these unchanged in construction of the orthonormal basis. Then $\dim W = r$ and $\dim V = r + m$. We shall prove that the set $\{u_1, \ldots, u_m\}$ is an orthonormal basis for $W^\perp$, thereby verifying the theorem.

Let $U$ be the vector space of dimension $m$ having as a basis the orthonormal set $\{u_1, \ldots, u_m\}$. If $u \in U$ then it can be expressed $\sum_{i=1}^{m} a_i u_i$. We then get

$$\langle u, w_j \rangle = \left\langle \sum_{i=1}^{m} a_i u_i, w_j \right\rangle = \sum_{i=1}^{m} a_i \langle u_i, w_j \rangle = 0,$$

for every $w_j$, element of the orthonormal basis of $W$, implying that $u \in W^\perp$. Hence $U \subseteq W^\perp$.

Next consider an arbitrary element $w$ of $W^\perp$. Then $w \in V$ and $w$ can be expressed

$$w = \sum_{i=1}^{n} b_i w_i + \sum_{i=1}^{m} c_i u_i.$$

This gives

$$\langle w, w_j \rangle = \left\langle \sum_{i=1}^{n} b_i w_i + \sum_{i=1}^{m} c_i u_i, w_j \right\rangle = \left\langle \sum_{i=1}^{n} b_i w_i, w_j \right\rangle + \left\langle \sum_{i=1}^{m} c_i u_i, w_j \right\rangle$$

$$= \sum_{i=1}^{n} b_i \langle w_i, w_j \rangle + \sum_{i=1}^{m} c_i \langle u_i, w_j \rangle = b_j,$$

since $\{w_1, \ldots, w_r\}$ is an orthonormal set and the set $\{u_1, \ldots, u_m\}$ is orthogonal to the set $\{w_1, \ldots, w_r\}$. But $\langle w, w_j \rangle$ is identically zero for each base element $w_j$ of $W$ since $w \in W^\perp$. Hence $b_j = 0$ for all $j = 1, \ldots, n$. Therefore

$$w = \sum_{i=1}^{m} c_i u_i,$$

implying that $w \in U$ and consequently that $W^\perp \subseteq U$.

Hence $W^\perp = U$.

The set $\{u_1, \ldots, u_m\}$ is therefore an orthonormal basis for $W^\perp$ and $\dim W^\perp = m$, giving $\dim V = \dim W + \dim W^\perp$.

*Example 2*

Let $\mathbf{R}^3$ be the normed inner product space of example 1 of this section and $W$ the subspace generated by the vectors (1, 0, 0) and (0, 1, 0). Then $W^\perp$ will be the subspace generated by (0, 0, 1). It can be seen that the dimension of $\mathbf{R}^3$ is 3, the dimension of $W$, 2, and the dimension of $W^\perp$, 1, illustrating the theorem.

*Theorem 2*

Let $V$ be a normed inner product space over the field of real numbers, having a positive definite inner product. If $W$ is a subspace of $V$, then

$$V = W \oplus W^\perp .$$

Proof: By the proof of the previous theorem, since the set

$$\{w_1, \ldots, w_r, u_1, \ldots, u_m\}$$

is a basis for $V$, the set $\{w_1, \ldots, w_r\}$ is a basis for $W$ and $\{u_1, \ldots, u_m\}$ is a basis for $W^\perp$,

$$V = W + W^\perp .$$

Let $w \in W \cap W^\perp$.

Then since $w$ is an element of $W$, it is orthogonal to every element of $W^\perp$. However since $w \in W \cap W^\perp$ it is an element of $W^\perp$. Hence

$$\langle w, w \rangle = 0 .$$

By the positive definiteness of the inner product, $w$ is thus the zero vector and $W \cap W^\perp = \{O\}$.

Thus $V = W + W^\perp$ and $W \cap W^\perp = \{O\}$, proving that $V = W \oplus W^\perp$.

## Exercises

1 Construct an orthonormal basis for the normed inner product space $\mathbf{R}^2$ from the set $\{(1, 2), (2, 1)\}$. Here the inner product is the dot product and the norm is the induced norm.

2 Construct an orthonormal basis for the inner product space $\mathbf{R}^3$ from the set $\{(2, 1, 0), (1, 3, 0), (0, 0, 2)\}$. The inner product is the usual dot product and the norm is the induced norm.

3 Let $V$ be the normed inner product space of continuous functions generated by $1, t, t^2$, over the interval [0, 1], with inner product defined by $\langle f, g \rangle = \int_0^1 fg\,dt$ and usual norm. Construct an orthonormal basis for this space.

4 Let $V$ be the vector space of continuous real-valued functions on the interval $[-1, 1]$ with inner product defined by

$$\langle f, g \rangle = \int_{-1}^{+1} f(x)\, g(x)\, dx .$$

Find the projection of the vector $x^3$ in the direction of $x$.

## 4.5 Dual Space, Functionals

Let $V$ be a finite-dimensional vector space over a field $\mathbf{F}$. Interpreting $\mathbf{F}$ as a vector space over itself, linear mappings from $V$ to $\mathbf{F}$ exist, and as has been previously shown this set of linear mappings can be made into a vector space over $\mathbf{F}$. Hence associated with every vector space $V$ is another vector space, denoted $V^*$, of linear mappings into the field of scalars. This vector space $V^*$ is called the *dual of V*. Its elements are called *functionals.*

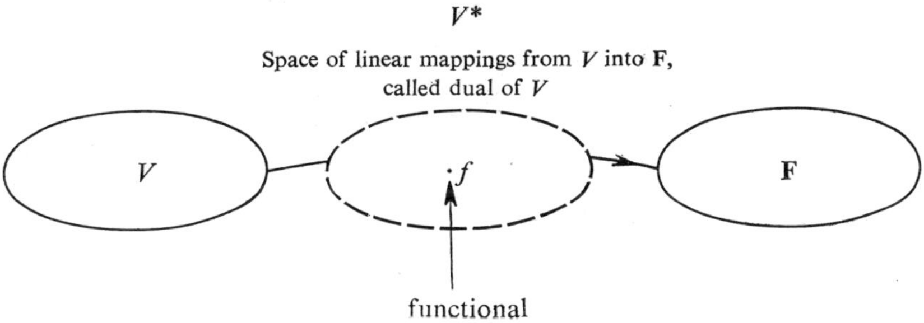


In this section and the next we shall be interested in the relative structures of the vector space $V$ and its dual $V^*$.[1]

*Theorem 1*

*Dimension V = Dimension V*.*
Proof: Let $\{v_1, \ldots, v_n\}$ be a basis for $V$ and $\{f_1, \ldots, f_n\}$ a set of elements in $V^*$ defined by

$$f_i(v_j) = \delta_{ij}.$$

By theorem 2, section 3.4 we know that such a set of functionals exists and that this definition determines each functional $f_i$ uniquely since its value on each base vector $v_j$ is given. For example $f_2$ will be the functional that maps the second base vector $v_2$ into the scalar 1, $f_2(v_2) = 1$, and any other base vector into zero, $f_2(v_k) = o$, $k \neq 2$. For $v$ an arbitrary vector of $V$, $v = \sum_{i=1}^{n} a_i v_i$, $f_2(v) = f_2\left(\sum_{i=1}^{n} a_i v_i\right) = \sum_{i=1}^{n} a_i f_2(v_i) = a_2$. Hence $f_2$ is defined. We shall show that this set $\{f_i, \ldots, f_n\}$ is a basis for $V^*$, proving the theorem.

Let $f$ be an arbitrary element of $V^*$ with values $c_1, \ldots, c_n$ on the base vectors of $V$, $f(v_i) = c_i$. Since $f$ is a linear mapping, the set $\{c_1, \ldots, c_n\}$ completely defines $f$. Using the properties of the Kronecker delta,

$$f(v_i) = c_i = \sum_{j=1}^{n} c_j \delta_{ji} = \sum_{j=1}^{n} c_j f_j(v_i).$$

1. To see the importance of the concept of the dual space in the application of linear algebra see, for example; D. G. Luenberger, *Optimization by Vector Space Methods,* (J. Wiley & Sons, Inc. 1969)

But since $V^*$ is a vector space, $\sum_{j=1}^{n} c_j f_j$, being a linear combination of the elements $f_1, \ldots, f_n$ of $V^*$, is itself an element of $V^*$ and hence is a functional. The above result tells us that the value of this functional $\sum_{j=1}^{n} c_j f_j$ on each base vector of $V$ is the same as the value of $f$ on that base vector. Hence $f$ and $\sum_{j=1}^{n} c_j f_j$ are the same functional,

$$f = \sum_{j=1}^{n} c_j f_j.$$

$f$ can therefore be expressed as a linear combination of the set $\{f_1, \ldots, f_n\}$, proving that the set spans $V^*$.

It remains to demonstrate that this set is linearly independent. Consider the identity

$$\sum_{i=1}^{n} a_i f_i = O,$$

for arbitrary scalars $a_1, \ldots, a_n$. Note that since $\sum_{i=1}^{n} a_i f_i$ is a linear combination of $f_1, \ldots, f_n$ it is itself a functional and an element of $V^*$. The zero on the right hand side of this equation is thus the zero functional, the functional that maps every element of $V$ onto the zero element of $\mathbf{F}$.

Operating on $v_j$ we have

$$\sum_{i=1}^{n} a_i f_i(v_j) = O(v_j),$$

giving

$$\sum_{i=1}^{n} a_i \delta_{ij} = o,$$

$$a_j = o, \quad \text{for all} \quad j = 1, \ldots, n.$$

Hence the set $\{f_1, \ldots, f_n\}$ is linearly independent and forms a basis for $V^*$, proving that the dimension of $V^*$ is $n$.

This basis is called the *dual basis* of the basis $\{v_1, \ldots, v_n\}$. There will exist such a dual basis for each basis of $V$.

*Example 1*

$\mathbf{R}^3$ is a vector space, hence it has a dual space $(\mathbf{R}^3)^*$. The set

$$\{(1, -1, 0), (1, 1, 0), (0, 0, 1)\}$$

is a basis for $\mathbf{R}^3$ find the dual basis.

Using the previous notation let $v_1 = (1, -1, 0)$, $v_2 = (1, 1, 0)$, $v_3 = (0, 0, 1)$. Then the dual basis is defined by $(f_1, f_2, f_3)$ where

$$f_i(v_j) = \delta_{ij}.$$

This gives $f_1$ to be the functional having the following values on the base vectors of $\mathbf{R}^3$:

$$f_1((1, -1, 0)) = 1,$$
$$f_1((1, 1, 0)) = 0,$$
$$f_1((0, 0, 1)) = 0.$$

Its value on an arbitrary element of $\mathbf{R}^3$, $(a, b, c)$ is given by

$$\begin{aligned} f_1((a, b, c)) &= f_1\left(\frac{a-b}{2}(1, -1, 0) + \frac{a+b}{2}(1, 1, 1) + c(0, 0, 1)\right) \\ &= \frac{a-b}{2} f_1((1, -1, 0)) + \frac{a+b}{2} f_1((1, 1, 0)) + c f_1((0, 0, 1)) \\ &= \frac{a-b}{2}. \end{aligned}$$

$f_1$ is defined either by its values on all the base vectors or by its value on an arbitrary element. Similarly the values of $f_2$ and $f_3$ on the base vectors of $\mathbf{R}^3$ can be written down and their effect on an arbitrary element of $\mathbf{R}^3$ found. It will be found that $f_2((a, b, c)) = (a + b)/2$ and that $f_3((a, b, c)) = c$. The dual basis $\{f_1, f_2, f_3\}$ is now known.

It should be noted that $V$ can be interpreted to be the space of functionals on $V^*$, defining the operation of $v \in V$ on $f \in V^*$ by

$$v(f) = f(v).$$

We then have that $V = (V^*)^*$. We write this $V = V^{**}$. The student is asked to prove in the exercises that the operation of $v$ on an element of $f$ thus defined gives $v$ to be a functional on $V^*$, and that every functional on $V^*$ can be thus represented.

### Exercises

1 The set $\{(1, 1), (2, 3)\}$ is a basis for $\mathbf{R}^2$; find the dual basis. Define it by expressing each element as a functional on each base vector of $\mathbf{R}^2$.

2 The set $\{(1, 0, 0), (0, 1, 0), (0, 0, 1)\}$ is a basis for $\mathbf{R}^3$. Find the dual basis in $(\mathbf{R}^3)^*$. Define it by expressing each element as a functional on an arbitrary element of $\mathbf{R}^3$.

3 The set $\{(1, 1, 0), (2, 1, 0), (0, 0, 1)\}$ is a basis for $\mathbf{R}^3$. Find the dual basis in $(\mathbf{R}^3)^*$. Define it by expressing each element as a functional on an arbitrary element of $\mathbf{R}^3$.

4 Let $V$ be the real two-dimensional vector space of functions defined on the interval $[0, 1]$ generated by $e^t$ and $e^{2t}$. Define the dual space and find the dual basis. Define the dual basis by expressing each element of the dual basis as a functional on an arbitrary element of $V$.

5 Let $V$ be the real two-dimensional vector space of functions defined on the interval $[-\pi/2, \pi/2]$ generated by $\cos t$, and $\sin t$. Define the dual space and find the dual basis. Define the dual basis by expressing each element of the dual basis as a functional on an arbitrary element of $V$.

6 Let $V$ be a vector space and $V^*$ its dual. Prove that $v \in V$ can be interpreted as a functional on $V^*$ if an operation is defined by $v(f) = f(v)$ for $f$ an arbitrary element of $V^*$. Prove that any functional on $V^*$ can be thus interpreted, proving that $V = (V^*)^*$.

## 4.6 The Reisz Representation Theorem and Natural Isomorphisms

In the previous section we defined the dual $V^*$ of a vector space $V$ and found that it was of the same dimension as the vector space. Since both vector spaces are over the same field of scalars, we know that they are isomorphic. We know that an isomorphism can be generated by putting a pair of bases of $V$ and $V^*$ in correspondence. For example, a basis for $V$ and its dual in $V^*$ could be used to define such a correspondence. All such isomorphisms are defined in terms of bases of $V$ and $V^*$. We shall demonstrate in this section that when $V$ is an inner product space an isomorphism can be defined between $V$ and its dual which is independent of bases. Such an isomorphism is called a *natural isomorphism*. Since in many situations we are interested in more than one basis for an inner product space, or in an arbitrary unspecified basis (as in tensor algebra, which we discuss in Chapter 7), this natural isomorphism which will be defined in terms of the inner product is of importance.

We now give the linear algebra version of the *Reisz Representation Theorem*; it leads to this natural isomorphism between $V$ and $V^*$. This is a very powerful theorem which the student will probably meet again, in a more general form, when he is introduced to the concept of the Hilbert Space[1] in more advanced courses.

*Theorem 1*

Let $V$ be a finite-dimensional inner product space over a field $\mathbf{F}$. For every element $v$ of $V$ there exists a functional $f_v$ such that $f_v(w) = \langle w, v\rangle$ for an arbitrary element $w$ *of* $V$.

Proof: Define a mapping $f_v$ of $V$ into $\mathbf{F}$ by $f_v(w) = \langle w, v\rangle$. We have to show that this mapping is linear.

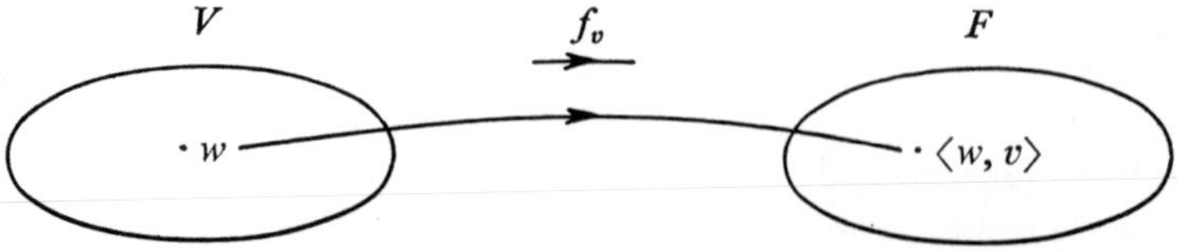


[1] See, for example, G. F. Simmons, *Topology and Modern Analysis* (McGraw-Hill, 1963).

For $w_1$ and $w_2$ elements of $V$, and arbitrary scalar $a$,

$$f_v(w_1 + w_2) = \langle w_1 + w_2, v\rangle = \langle w_1, v\rangle + \langle w_2, v\rangle = f_v(w_1) + f_v(w_2),$$

and

$$f_v(aw_1) = \langle aw_1, v\rangle = a\langle w_1, v\rangle = af_v(w_1).$$

Hence, $f_v$ is a functional.

This theorem defines a mapping $h$ of $V$ into $V^*$, $h: v \mapsto f_v$.

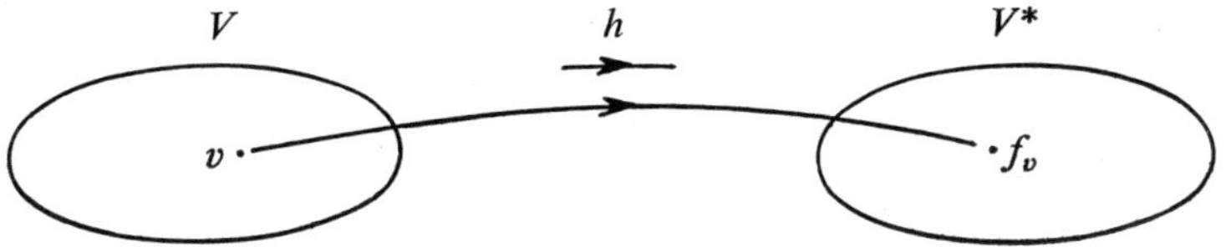


We aim now to prove that this mapping of $V$ into $V^*$ is an isomorphism. This shows that every element of $V^*$ arises in this manner.

*Theorem 2*

The mapping $h$ of $V$ into $V^*$ defined by $h(v) = f_v$, where $f_v(w) = \langle w, v\rangle$, $w \varepsilon V$, is an isomorphism. Thus not only does every vector in $V$ define a functional in this manner but the converse holds.

Proof: Let us first of all prove the linearity of $h$. For arbitrary elements $v_1$ and $v_2$ of $V$,

$$h(v_1) = f_{v_1}, \quad h(v_2) = f_{v_2} \quad \text{and} \quad h(v_1 + v_2) = f_{v_1+v_2}.$$

But

$$f_{v_1+v_2}(w) = \langle w, v_1 + v_2\rangle = \langle w, v_1\rangle + \langle w, v_2\rangle = f_{v_1}(w) + f_{v_2}(w),$$

for arbitrary $w \varepsilon V$, implying that

$$f_{v_1+v_2} = f_{v_1} + f_{v_2}.$$

Hence

$$h(v_1 + v_2) = h(v_1) + h(v_2),$$

proving the first part of the linearity of $h$.

For an arbitrary scalar $a$,

$$h(v_1) = f_{v_1}, \quad \text{and} \quad h(av_1) = f_{av_1}.$$

We have that

$$f_{av_1}(w) = \langle w, av_1\rangle = a\langle w, v_1\rangle = af_{v_1}(w),$$

for arbitrary $w \varepsilon V$, implying that $f_{av_1} = af_{v_1}$.

Hence

$$h(av_1) = ah(v_1)$$

completing the proof of the linearity of $h$. The range of $h$ is thus a subspace of $V^*$.

Let us next prove that $h$ is injective by showing that the kernel of the mapping is the zero element of $V$. We have that $v$ is an element of the kernel of $h$ if and only if $h(v) = O$, the zero functional. However, if the image of $v$ under $h$, $f_v$, is the zero functional, $f_v(w) = O$ for all $w \varepsilon V$. But this implies that $\langle w, v \rangle = o$ for all $w \varepsilon V$, giving, by the non-degeneracy of the inner product, that $v = O$. Hence $h$ is injective.

Finally, since the dimension of the domain is equal to the dimension of the range plus the dimension of the kernel, and since the dimension of the kernel is zero, the dimension of the range must be equal to the dimension of $V$. Hence the dimension of the range must be equal to the dimension of $V^*$, and since the range is a subspace of $V^*$ therefore the range is $V^*$, proving that $h$ is surjective.

The mapping is thus an isomorphism.

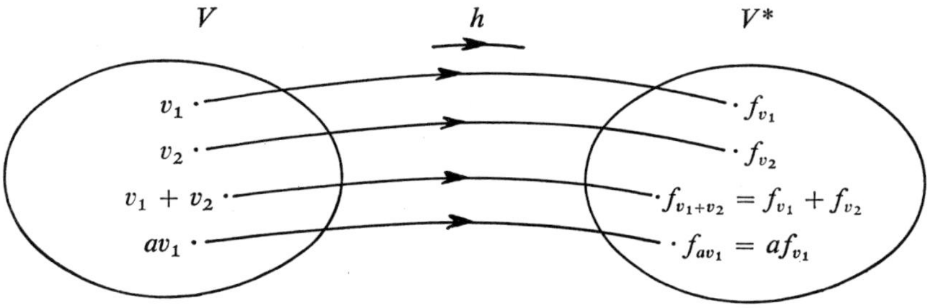


This isomorphism has been defined using the inner product of $V$ and is independent of any basis selected to discuss $V$. It is an isomorphism defined in terms of the intrinsic properties of the inner product space rather than in terms of a special basis; for this reason it is called a *natural isomorphism.*

*Example 1*

Let $\mathbf{R}^2$ be the real inner product space with inner product defined by the dot product. Let $h$ be the natural isomorphism between $\mathbf{R}^2$ and $(\mathbf{R}^2)^*$ determined by this inner product. Define this isomorphism.

The isomorphism, like any linear mapping, will be defined if its effect on each base vector is known, for any convenient basis. Let us choose as a basis for $\mathbf{R}^2$ the usual basis $\{(1, 0), (0, 1)\}$. Then

$$h: (1, 0) \mapsto f_{(1,0)},$$

and

$$(0, 1) \mapsto f_{(0,1)}.$$

We have to determine $f_{(1,0)}$ and $f_{(0,1)}$.

$f_{(1,0)}$ will be the element of $(\mathbf{R}^2)^*$ that is defined by

$$f_{(1,0)}(a, b) = \langle (a, b), (1, 0) \rangle = (a, b) \cdot (1, 0)$$
$$= a.$$

Similarly, $f_{(0,1)}$ is the element of $(\mathbf{R}^2)^*$ defined by

$$f_{(0,1)}(a, b) = \langle (a, b), (0, 1) \rangle = (a, b) \cdot (0, 1).$$
$$= b.$$

Knowing the effect of $h$ on the base elements $(1, 0)$ and $(0, 1)$ of $\mathbf{R}^2$ defines $h$ completely.

However, an alternative and actually more useful way of defining $h$ is to give the image of an arbitrary element of $\mathbf{R}^2$ under $h$. That is, if $(p, q) \in \mathbf{R}^2$ and

$$h: (p, q) \mapsto f_{(p,q)},$$

to give $f_{(p,q)}$.

$f_{(p,q)}$ is the element of $(\mathbf{R}^2)^*$ defined by

$$f_{(p,q)}(a, b) = \langle (a, b), (p, q) \rangle = (a, b) \cdot (p, q)$$
$$= ap + bq.$$

Hence $h$ is the isomorphism of $\mathbf{R}_2$ onto $(\mathbf{R}_2)^*$ defined by

$$h: (p, q) \mapsto f_{(p,q)},$$

where $f_{(p,q)}(a, b) = ap + bq$.

The image of the element $(1, 3)$ under this isomorphism, for example, is the functional $f_{(1,3)}$ defined by $f_{(1,3)}(a, b) = a + 3b$.

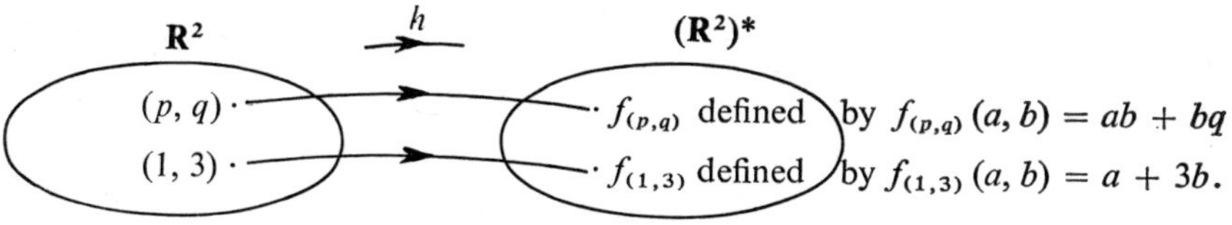


The discussions of this section and the previous one can now be combined to give an interesting interpretation of the representation of an arbitrary vector relative to a chosen basis. Let $v$ be an arbitrary element of an inner product space $V$, $\{v_1, \ldots, v_n\}$ a chosen basis for $V$ and $\{f_1, \ldots, f_n\}$ the dual basis in $V^*$. We know that there exist scalars $a_1, \ldots, a_n$ such that $v$ can be represented

$$v = a_1 v_1 + \cdots + a_n v_n.$$

However, we have not given any interpretation to the scalars $a_1, \ldots, a_n$. We are now in a position to do so. Let $h$ be the natural isomorphism of $V$ onto $V^*$ defined by the inner product. Let $h^{-1}f_i = u_i$. The set $\{u_1, \ldots, u_n\}$ is a set of $n$ linearly independent elements in $V$ (theorem 5, section 3.4) and hence forms a basis for $V$. Since $h$ is an isomorphism the spaces $V$ and $V^*$ are considered as identical mathematically under the correspondence defined by $h$.

The set $\{f_1, \ldots, f_n\}$ is called the dual basis of $\{v_1, \ldots, v_n\}$ in $V^*$. We call the basis $\{u_1, \ldots, u_n\}$ the *dual* of $\{v_1, \ldots, v_n\}$ in $V$, since it is the image of the set $\{f_1, \ldots, f_n\}$ under $h^{-1}$.

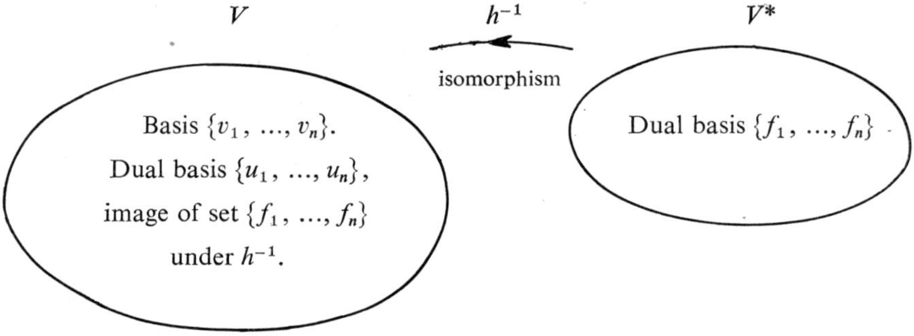


Since the set $\{f_1, \ldots, f_n\}$ is defined by $f_i(v_j) = \delta_{ij}$ and since $u_i$ is defined by $\langle w, u_i \rangle = f_i(w)$, we have,

$$\langle v_j, u_i \rangle = f_i(v_j) = \delta_{ij}.$$

Hence the dual basis of $\{v_1, \ldots, v_n\}$ in $V$ is defined by

$$\langle v_j, u_i \rangle = \delta_{ij}.$$

Returning now to the consideration of the representation of $v$ relative to the basis $\{v_1, \ldots, v_n\}$, $v = a_1 v_1 + \cdots + a_n v_n$, we have on taking the inner product with $u_i$,

$$\begin{aligned} \langle v, u_i \rangle &= \langle a_1 v_1 + \cdots + a_n v_n, u_i \rangle \\ &= \langle a_1 v_1, u_i \rangle + \cdots + \langle a_n v_n, u_i \rangle \\ &= a_1 \delta_{1i} + \cdots + a_n \delta_{ni} \\ &= a_i, \end{aligned}$$

giving that $a_i = \langle v, u_i \rangle$.

Hence the representation of $v$ relative fo the basis $\{v_1, \ldots, v_n\}$ is

$$\underline{v = \langle v, u_1 \rangle\, v_1 + \cdots + \langle v, u_n \rangle\, v_n,}$$

or alternatively,

$$\underline{v = f_1(v)\, v_1 + \cdots + f_n(v)\, v_n,}$$

where the sets $\{u_1, \ldots, u_n\}$ and $\{f_1, \ldots, f_n\}$ are the duals of the set $\{v_1, \ldots, v_n\}$ in $V$ and $V^*$ respectively. This result is actually of more theoretical importance than of practical importance in finding the representation.

It can now be seen that the representation mapping of $V$ onto $\mathbf{F}^n$ is defined by

$$v \mapsto (\langle v, u_1\rangle, \ldots, \langle v, u_n\rangle),$$

or alternatively

$$v \mapsto (f_1(v), \ldots, f_n(v)).$$

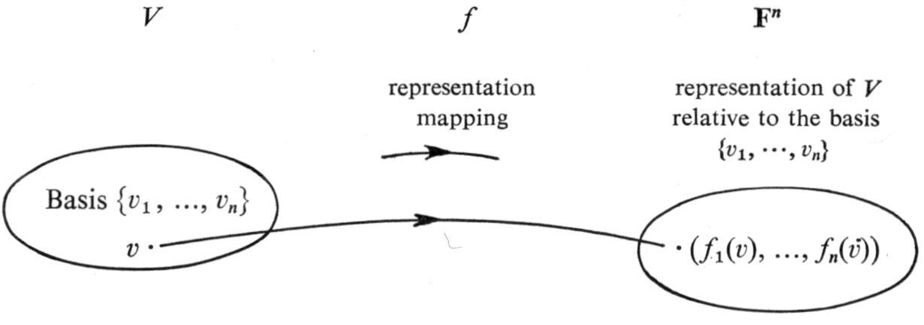


If the vector space is $\mathbf{F}^n$ then the set $\{f_1, \ldots, f_n\}$ which is the dual of the usual basis $\{(1, 0 \ldots, 0), \ldots, (0, \ldots, 0, 1)\}$, where 1 is the multiplicative identity of the field $\mathbf{F}$, is usually called the *set of slot functions* on $\mathbf{F}^n$. In such a case, if $v = (a_1, \ldots, a_n)$ is an arbitrary element of $\mathbf{F}^n$ then by the above theory $v = (f_1(v), \ldots, f_n(v))$ and $f_i(v) = a_i$.

*Example 2*

Let $\mathbf{R}^3$ be the real inner product space with inner product defined as dot product. The set $\{(1, 1, 0), (1, -1, 0), (0, 0, 1)\}$ is a basis for $\mathbf{R}^3$.

Find the dual basis in $\mathbf{R}^3$.

Let $\{(a, b, c), (p, q, r), (l, m, n)\}$ be the dual basis. Then $(a, b, c)$ is defined by $\langle(1, 1, 0), (a, b, c)\rangle = 1$, $\langle(1, -1, 0), (a, b, c)\rangle = 0$, $\langle(0, 0, 1), (a, b, c)\rangle = 0$. Using the dot product for inner product, these conditions reduce to

$$a + b = 1, \ a - b = 0, \ c = 0,$$

giving the first vector in the dual basis to be $(\frac{1}{2}, \frac{1}{2}, 0)$.

Similarly $(p, q, r)$ is defined by $\langle(1, 1, 0), (p, q, r)\rangle = 0$, $\langle(1, -1, 0), (p, q, r)\rangle = 1$, $\langle(0, 0, 1), (p, q, r)\rangle = 0$. These conditions, using the dot product, give $(p, q, r)$ to be the vector $(\frac{1}{2}, -\frac{1}{2}, 0)$. The vector $(l, m, n)$ must satisfy the conditions $\langle(1, 1, 0), (l, m, n)\rangle = 0$, $\langle(1, -1, 0), (l, m, n)\rangle = 0$, $\langle(0, 0, 1), (l, m, n)\rangle = 1$ implying that $(l, m, n)$ is the vector $(0, 0, 1)$.

Hence the dual basis is $\{(\frac{1}{2}, \frac{1}{2}, 0), (\frac{1}{2}, -\frac{1}{2}, 0), (0, 0, 1)\}$.

If the dual of a basis turns out to be the original set the basis is said to be *self-dual*. The student is asked to prove in the exercises that a necessary and sufficient condition for a basis to be self-dual is that it be orthonormal.

## Exercises

1 Let $\mathbf{R}^2$ be the real inner product space with inner product defined by the dot product. Let $h$ be the natural isomorphism between $\mathbf{R}^2$ and $(\mathbf{R}^2)^*$ defined by $v \mapsto f_v$, where $f_v(w) = \langle w, v\rangle$, for $w \in \mathbf{R}^2$. Find the image of the element (1, 1) of $\mathbf{R}^2$ under this isomorphism.

2 Let $\mathbf{R}^3$ be the real inner product space with inner product defined by $\langle(a, b, c), (p, q, r)\rangle = 3ap + 3bq + 2cr$, i.e. the inner product space of exercise 6, section 4.1. Let $h$ be the natural isomorphism of $\mathbf{R}^3$ onto $(\mathbf{R}^3)^*$ defined by this inner product. Define this isomorphism by finding the images of base elements of $\mathbf{R}^3$ in $(\mathbf{R}^3)^*$. You may select any basis you choose.

3 Let $V$ be the real inner product space of continuous functions on the interval [0, 1] spanned by $t$ and $t^2$, with inner product defined by

$$\langle f, g\rangle = \int_0^1 fg\, dt.$$

Let $h$ be the natural isomorphism of $V$ onto $V^*$ defined by this inner product. Define this isomorphism by finding the image of an arbitrary element of $V$ in $V^*$.

4 Let $V$ be the real inner product space of continuous functions on the interval $[-\pi/2, \pi/2]$ spanned by $\cos t$ and $\sin t$ with inner product defined by

$$\langle f, g\rangle = \int_{-\pi/2}^{\pi/2} fg\, dt.$$

Let $h$ be the natural isomorphism of $V$ onto $V^*$ defined by this inner product. Find the image of the element $\cos t + 2 \sin t$ of $V$ in $V^*$ under this isomorphism.

5 Let $\mathbf{R}^3$ be the real inner product space with inner product defined by the dot product. Find the dual of the basis {(1, 2, 0), (2, 1, 0), (0, 1, 0)} in $\mathbf{R}^3$.

6 Let $\mathbf{R}^3$ be the real inner product space with inner product defined by the dot product and usual norm. Show that the basis {(1, 0, 0), (0, 1, 0), (0, 0, 1)} is self-dual.

7 Show that a necessary and sufficient condition for a basis of an inner product space to be self-dual is that it be orthonormal.

## 4.7 Hermitian Product

Mathematicians have found that a *Hermitian product* is a more interesting and useful structure to study than an inner product on a complex vector space. This is a mapping that assigns to any pair of vectors $u_1$ and $u_2$ of a complex vector space a number $\langle u_1, u_2\rangle$ satisfying the following axioms:

1 $\langle u_1, u_2\rangle = \overline{\langle u_2, u_1\rangle}$ for any pair of vectors $u_1$ and $u_2$ of the complex vector space $U$. Here the bar denotes complex conjugate. This property is called the *Hermitian symmetry property*.

2 $\langle u_1, u_2 + u_3\rangle = \langle u_1, u_2\rangle + \langle u_1, u_3\rangle$ for any triple of elements $u_1$, $u_2$ and $u_3$ of $U$.

3 If $a$ is a complex number,

$$\langle au_1, u_2\rangle = a\langle u_1, u_2\rangle \quad \text{and} \quad (u_1, au_2\rangle = \bar{a}\langle u_1, u_2\rangle,$$

for any pair of vectors $u_1$, $u_2$ of $U$.

A Hermitian product is said to be *positive definite* if $\langle u, u\rangle \geqq 0$ for every $u \varepsilon U$ and $\langle u, u\rangle = 0$ if and only if $u$ is the zero vector. A complex vector space on which a positive-definite Hermitian product has been defined is called a *unitary space*. In an analogous manner to the inner product discussion we shall define a Hermitian product to be *non-degenerate* if for $u \varepsilon U$, $\langle u, w\rangle = 0$ for every $w \varepsilon U$ implies that $u$ is the zero vector. In this section we shall be interested in positive definite, non-degenerate Hermitian products.

The properties of a Hermitian product defined on a complex vector space are analogous to those of an inner product defined on a real vector space, and the proofs of these properties are similar to the proofs of the properties of an inner product.

*Example 1*

Consider the complex vector space $\mathbf{C}^n$ (the set of $n$-tuples of complex numbers with the natural vector-space structure defined in Section 1.4). If $X = (x_1, \ldots, x_n)$ and $Y = (y_1, \ldots, y_n)$ are two elements of $\mathbf{C}^n$ define their Hermitian product:

$$\langle X, Y\rangle = x_1\bar{y}_1 + \cdots + x_n\bar{y}_n .$$

As can be seen, the axioms of the Hermitian product are satisfied. This is a positive-definite Hermitian product, called the *usual Hermitian product* for $\mathbf{C}^n$.

*Example 2*

Let $V$ be the complex vector space of complex-valued functions on the interval [0, 1]. The operations of addition and multiplication are defined pointwise. If $f, g \varepsilon V$ then a positive definite Hermitian product on $V$ is defined by

$$\langle f, g\rangle = \int_0^1 f(t)\, \bar{g}(t)\, dt .$$

A positive definite Hermitian product can be used in exactly the same way as a positive definite inner product to *generate a norm* on the space. If $v \varepsilon V$, the complex vector space, and $\langle\, , \rangle$ is the Hermitian product, then the norm generated by the Hermitian product is defined by

$$\|v\| = \sqrt{\langle v, v\rangle} .$$

*Example 3*

Let $\mathbf{C}^3$ have the usual vector space structure with the usual Hermitian product. If $X = (x_1, x_2, x_3)$ and $Y = (y_1, y_2, y_3) \varepsilon \mathbf{C}^3$ then

$$\langle X, Y\rangle = x_1\bar{y}_1 + x_2\bar{y}_2 + x_3\bar{y}_3 .$$

Since this Hermitian product is positive definite it can be used to generate the following norm on the space

$$\|X\| = \sqrt{x_1\bar{x}_1 + x_2\bar{x}_2 + x_3\bar{x}_3}.$$

This example illustrates why this Hermitian product is more useful on $\mathbf{C}^3$ than is the usual inner product (dot product). The Hermitian product is positive definite, whereas the inner product is not. The inner product cannot be used to generate a norm.

We now list further results for complex vector spaces having Hermitian products. The student is asked to satisfy himself of the validity of these results in the exercises.

We can define *orthogonality*, *orthogonal complement*, *orthogonal basis* and *projection* in terms of a Hermitian product similarly to the way we defined them in terms of an inner product. The *Schwartz inequality* holds for a positive definite Hermitian product, the proof of this result being similar to the case of the inner product. A *Gram-Schmidt orthogonalization process* can be used to construct an orthonormal basis for a complex vector space which has a positive definite Hermitian product in the same way as for the inner product space. Here orthogonality and norm are defined using the Hermitian product. Theorems 1 and 2 of section 4.4 both have natural analogs for a complex vector space with Hermitian product.

As in theorem 1 of section 4.6, the mapping $v \mapsto f_v$, where $f_v(w) = \langle w, v\rangle$, defines a linear functional $f_v$ on $V$ and hence is a mapping of $V$ into $V^*$. It can be shown as in section 4.6 that this mapping $h$ is bijective. It is not, however, an isomorphism. The student is asked to prove that if $h(v) = f_v$ then $h(av) = \bar{a}f_v$, illustrating that the mapping $h$ is not linear here. However, this disadvantage of the Hermitian product is outweighed by the advantage that it can be used to generate a norm in a natural manner for many complex vector spaces.

Note that the axioms of a Hermitian product are satisfied by an inner product defined on a real vector space. The definition of the Hermitian product can thus be extended to real vector spaces; on such spaces the two concepts are identical.

## Exercises

1 Prove that the Hermitian products defined in both example 1 and example 2 of this section are positive definite Hermitian products which are non-degenerate.

2 Use a Hermitian product defined on a complex vector space $V$ to define orthogonality, orthogonal complement, orthogonal basis and projection on the space.

3 Prove the Schwartz inequality for a Hermitian product on a complex vector space.

4 Let $V$ be a complex vector space with Hermitian product. Prove that the mapping $h$ of $V$ into $V^*$ defined by $h\colon v \mapsto f_v$, where $f_v(w) = \langle w, v \rangle$ has the properties

$$h(v_1 + v_2) = h(v_1) + h(v_2) \quad \text{and} \quad h(av) = \bar{a}h(v).$$

This last property shows that the mapping $h$ is not a linear mapping as is the case with an inner product. Hence the Hermitian product cannot be used to define a natural isomorphism between a complex vector space and its dual.

CHAPTER 5

# Determinants, further properties of Matrices and Systems of Linear Equations

## 5.1 Determinants

Let $M_{\mathbf{F}}$ be the set of square matrices over a field $\mathbf{F}$. The *determinant* is a special mapping of this set into $\mathbf{F}$. That is, the determinant associates with every square matrix an element of $\mathbf{F}$, called the determinant of the matrix. If $D$ denotes the mapping then the determinant of a square matrix $A$ is $D(A)$.

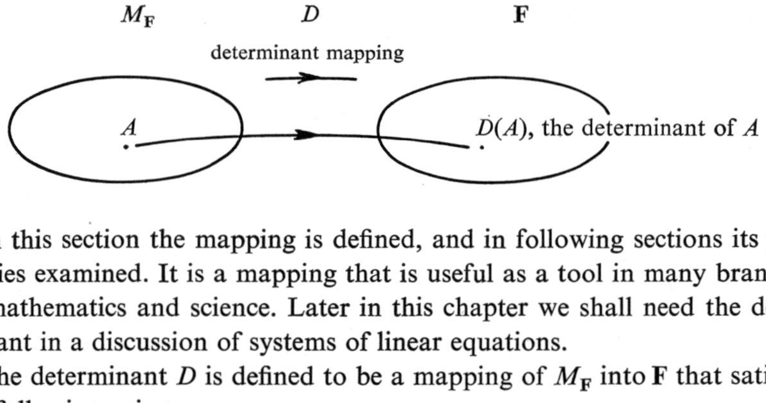


In this section the mapping is defined, and in following sections its properties examined. It is a mapping that is useful as a tool in many branches of mathematics and science. Later in this chapter we shall need the determinant in a discussion of systems of linear equations.

The determinant $D$ is defined to be a mapping of $M_{\mathbf{F}}$ into $\mathbf{F}$ that satisfies the following axioms:

1 If $D$, mapping an element $A$ of $M_{\mathbf{F}}$ into $\mathbf{F}$, is interpreted as a mapping of the columns of $A$ into $\mathbf{F}$, it is a skew-symmetric multilinear mapping in these columns.

That is, if the columns of an $n \times n$ matrix $A$ are $A^1, \ldots, A^n$ then

$$D(A) = D(A^1, \ldots, A^i, \ldots, A^j, \ldots, A^n) = -D(A^1, \ldots, A^j, \ldots, A^i, \ldots, A^n),$$

and

$$D(A^1, \ldots, A^i + B^i, \ldots, A^n) = D(A^1, \ldots, A^i, \ldots, A^n) + (A^1, \ldots, B^i, \ldots, A^n),$$

$$D(A^1, \ldots, cA^i, \ldots, A^n) = cD(A^1, \ldots, A^i, \ldots, A^n).$$

2 If $I$ is a unit matrix then $D(I) = 1$, the multiplicative identity of the field.

For the time being we assume the existence of the determinant.
The determinant of $(a_{ij})$ is usually denoted

$$\begin{vmatrix} a_{11} & \dots & a_{1n} \\ \vdots & & \\ a_{n1} & \dots & a_{nn} \end{vmatrix}.$$

*Examples to illustrate the properties of determinants*

Consider the set of square matrices over $\mathbf{R}$.

1 $\begin{vmatrix} 1 & 2 \\ 3 & 4 \end{vmatrix} = -\begin{vmatrix} 2 & 1 \\ 4 & 3 \end{vmatrix}$, by the skew symmetry of the first axiom.

2 $\begin{vmatrix} 1 & 2 & 3 \\ 4 & 5 & 6 \\ 7 & 8 & 9 \end{vmatrix} = -\begin{vmatrix} 2 & 1 & 3 \\ 5 & 4 & 6 \\ 8 & 7 & 9 \end{vmatrix} = +\begin{vmatrix} 2 & 3 & 1 \\ 5 & 6 & 4 \\ 8 & 9 & 7 \end{vmatrix}$, again by the skew symmetry of the first axiom.

3 $3\begin{vmatrix} 1 & 2 \\ 3 & 4 \end{vmatrix} = \begin{vmatrix} 3 & 2 \\ 9 & 4 \end{vmatrix} = \begin{vmatrix} 1 & 6 \\ 3 & 12 \end{vmatrix}$, by the linearity of $D$ in the columns of the matrix.

4 $\begin{vmatrix} 1 & (3+4) & 1 \\ 2 & (2+3) & 0 \\ 3 & (1+2) & 2 \end{vmatrix} = \begin{vmatrix} 1 & 3 & 1 \\ 2 & 2 & 0 \\ 3 & 1 & 2 \end{vmatrix} + \begin{vmatrix} 1 & 4 & 1 \\ 2 & 3 & 0 \\ 3 & 2 & 2 \end{vmatrix}$, again by the linearity of $D$ in the columns of the matrix.

5 $\begin{vmatrix} 1 & 0 & 0 \\ 0 & 1 & 0 \\ 0 & 0 & 1 \end{vmatrix} = 1$, and $\begin{vmatrix} 1 & 0 \\ 0 & 1 \end{vmatrix} = 1$, by the second axiom.

We shall prove that the previous axioms define the determinant uniquely. First we need two lemmas.

*Lemma 1*

If $A$ is a square matrix in which two columns are identical then $D(A) = o$.
Proof: Let the columns of $A$ be $A^1, \dots, A^n$ and let the ith column be identical to the $j$th column, $A^i = A^j$.

We then have that

$$A = (A^1 \dots A^i \dots A^j \dots A^n) = (A^1 \dots A^j \dots A^i \dots A^n),$$

interchanging the $i$th and the $j$th columns. This gives

$$D(A) = D(A^1, \dots, A^i, \dots, A^j, \dots, A^n) = +D(A^1, \dots, A^j, \dots, A^i, \dots, A^n).$$

But by the skew symmetry of the columns of $D(A)$,

$$D(A^1, \dots, A^i, \dots, A^j, \dots, A^n) = -D(A^1, \dots, A^j, \dots, A^i, \dots, A^n).$$

Hence $D(A) = o$.

Note that this property of $D$ follows from the skew-symmetry condition of the first axiom. Any mapping of $M_{\mathbf{F}}$ into $\mathbf{F}$ that satisfies this condition will have this property. We make use of this result in the proof of lemma 2.

*Example*

$$\begin{vmatrix} 1 & 0 & 1 \\ 2 & 1 & 2 \\ 3 & 2 & 3 \end{vmatrix} = 0, \text{ since the first and third columns are identical.}$$

*Lemma 2*

Let $G$ be a mapping of $M_{\mathbf{F}}$ into $\mathbf{F}$ satisfying the first axiom of the determinant. If $A \varepsilon M_{\mathbf{F}}$ then there exists a scalar $a$ such that $G(A) = aG(I)$, $I$ being the multiplicative identity matrix of the same kind as $A$.

Proof: Let $A$ be an $n \times n$ matrix and let $I_i$ denote the column vector with $n$ components, having zeros in all slots except the $i$th, in which it has 1. Then $A$ can be expressed

$$A = \left(\sum_{i=1}^{n} a_{i1} I_i, \ldots, \sum_{j=1}^{n} a_{jn} I_j\right),$$

since each column can be expressed as a linear combination of the set $\{I_1, \ldots, I_n\}$.

Hence $G(A) = G\left(\sum_{i=1}^{n} a_{i1} I_i, \ldots, \sum_{j=1}^{n} a_{jn} I_j\right)$.

Using the properties of $G$ defined by axiom 1 and the result of the discussion of lemma 1, this expression reduces to

$$G(A) = aG(I_1, \ldots, I_n) = aG(I),$$

where $a$ is a scalar that depends on the matrix $A$. (The reader is asked to verify this in the exercises that follow.)

*Theorem*

The axioms of a determinant define a unique mapping.

Proof: Suppose both $D$ and $D'$ satisfy the axioms of the determinant. Let $G = D - D'$ be a mapping of $M_{\mathbf{F}}$ into $\mathbf{F}$ defined by $G(A) = D(A) - D'(A)$. Then $G$ can be seen to satisfy the conditions of axiom 1. Hence by the above lemma there exists a scalar $a$ such that

$$G(A) = aG(I),$$

for arbitrary $A \varepsilon M_{\mathbf{F}}$.

However for the matrix $I$,

$$G(I) = D(I) - D'(I) = 1 - 1 = o.$$

The previous equation then implies that $G$ is the zero mapping of $M_{\mathbf{F}}$ into $\mathbf{F}$. Hence $D = D'$ and the determinant is uniquely determined by the axioms.

A square matrix $A$ is said to be *singular* if it has zero determinant, otherwise it is *non-singular*. We shall see the importance of non-singular matrices later in this chapter when we discuss the existence and uniqueness of solutions to systems of linear equations. It will also be seen that inverses under the operation of multiplication exist only for nonsingular matrices.

### Exercises

1 In the proof of Lemma 2 verify that $G\left(\sum_{i=1}^{n} a_{i1}I_i, \ldots, \sum_{j=1}^{n} a_{jn}I_j\right)$ reduces to $aG(I_1, \ldots, I_n)$.

2 In the proof of the uniqueness theorem for the determinant it was stated that if $D$ and $D'$ are mappings satisfying the axioms of the determinant then the mapping $D - D'$ satifies the first axiom of the determinant. Prove this result.

## 5.2 Evaluation of Determinants

Although we have uniquely defined the determinant function in the previous section we have not as yet given a method of evaluating the determinant of a square matrix. We shall give a method in this section.

Let $A = (a_{ij})$ be an $n \times n$ matrix. By deleting the $i$th row and $j$th column of $A$ we get an $n - 1 \times n - 1$ matrix which is called the *minor* of the element $a_{ij}$. There will exist such a minor for every element of the matrix. Denote the minor of the element $a_{ij}$ by $A_{ij}$.

*Example 1*

If $A = \begin{pmatrix} 1 & 0 & 2 \\ 3 & 4 & 1 \\ 2 & 4 & 1 \end{pmatrix}$ then $A_{11} = \begin{pmatrix} 4 & 1 \\ 4 & 1 \end{pmatrix}$, $A_{12} = \begin{pmatrix} 3 & 1 \\ 2 & 1 \end{pmatrix}$, $A_{23} = \begin{pmatrix} 1 & 0 \\ 2 & 4 \end{pmatrix}$, etc.

*Theorem*

Let $A$ be an $n \times n$ matrix, then the mapping $D$ defined by $D(A) = A$, for $n = 1$,

$$D(A) = \sum_{i=1}^{n} (-1)^{i+j} a_{ij} D(A_{ij}), \quad \text{for} \quad n > 1,$$

any $j = 1, \ldots, n$ is the determinant. This is called the expansion of the determinant in terms of the $j$th column. The determinant can be expanded thus in terms of any column, expressing the determinant of an $n \times n$ matrix as a linear combination of determinants of $n - 1 \times n - 1$ matrices.

Proof: It is necessary and sufficient to show that the mapping satisfies the axioms of the determinant. The uniqueness of the determinant mapping completes the proof. The mapping $D$ is here defined inductively in that $D(A_{ij})$ is the image of the $n - 1 \times n - 1$ matrix $A_{ij}$ under $D$.

It can be seen that $D$: $A \mapsto A$, $A$ a scalar, satisfies the axioms of the determinant. For $A$ a $2 \times 2$ matrix $\begin{pmatrix} a & b \\ c & d \end{pmatrix}$, this mapping gives, expanding in terms of the first column,

$$\begin{aligned} D(A) &= \sum_{i=1}^{2} (-1)^{i+j} a_{ij} D(A_{ij}) \\ &= (-1)^{1+1} a_{11} D(A_{11}) + (-1)^{2+1} a_{21} D(A_{21}) \\ &= ad - bc. \end{aligned}$$

The same result for $D(A)$ is obtained expanding in terms of the second column. It can easily be checked (left as an exercise) that all the axioms of the determinant are satisfied. Hence the mapping is the determinant mapping for $2 \times 2$ matrices. Let us assume that the mapping is the determinant mapping for $n - 1 \times n - 1$ matrices. We shall prove that it satisfies all the necessary axioms for $n \times n$ matrices thereby proving by mathematical induction that it is the determinant mapping for all square matrices.

Let $B = (b_{ij})$ be the matrix obtained from $A$ by interchanging the $h$ and $m$ columns of $A$. We shall show that $D(B) = -D(A)$. If $j \neq h$ or $m$, then $b_{ij} = a_{ij}$ and $D(B_{ij}) = -D(A_{ij})$ for all values of $i$, since $B_{ij}$ is obtained from $A_{ij}$ by interchanging the $h$ and $m$ columns. Hence

$$\sum_{i=1}^{n} (-1)^{i+j} b_{ij} D(B_{ij}) = -\sum_{i=1}^{n} (-1)^{i+j} a_{ij} D(A_{ij}).$$

We shall have to consider the cases of $j = h$ and $j = m$; that is when one of the columns interchanged is the column used for the expansion. Suppose $j = h$ and $m > h$, then we have to examine $\sum_{i=1}^{n} (-1)^{i+h} b_{ih} D(B_{ih})$. We can expand $D(B_{ih})$ using the same mapping, applied to $n - 1 \times n - 1$ matrices. Let the minor of the element $b_{km}$ in $B_{ih}$ be $B_{ihkm}$. We shall expand $D(B_{ih})$ in terms of its $(m - 1)$ column. Note that since $m > h$ then the elements that make up the $(m - 1)$ column of $B_{ih}$ lie in the $m$ column of $B$. Therefore

$$D(B_{ih}) = \sum_{k=1}^{i-1} (-1)^{k+(m-1)} b_{km} D(B_{ihkm}) + \sum_{k=i+1}^{n} (-1)^{k-1+(m-1)} b_{km} D(B_{ihkm}).$$

We have to break $D(B_{ih})$ up thus since the elements of the $k$ row of $B$ for $k > i$, lie in the $k - 1$ row of $B_{ih}$. Hence

$$\begin{aligned} \sum_{i=1}^{n} (-1)^{i+j} b_{ij} D(B_{ij}) = \sum_{i=1}^{n} (-1)^{i+h} b_{ih} \Bigg[ &\sum_{k=1}^{i-1} (-1)^{k+m-1} b_{km} D(B_{ihkm}) \\ &+ \sum_{k=i+1}^{n} (-1)^{k-1+(m-1)} b_{km} D(B_{ihkm}) \Bigg]. \end{aligned}$$

Further, $b_{ih} = a_{im}$ and $b_{km} = a_{kh}$ since the $h$ and $m$ columns of $A$ are interchanged to get $B$, and $B_{ihkm} = A_{imkh}$ since $B_{ihkm}$ is the minor of the element $b_{km}$ in $B_{ih}$ and $A_{imkh}$ is the minor of the element $a_{kh}$ in $A_{im}$. Hence

$$\sum_{i=1}^{n}(-1)^{i+j}\, b_{ij}D(B_{ij}) = \sum_{i=1}^{n}(-1)^{i+m}\, a_{im}\left[\sum_{k=1}^{i-1}(-1)^{k+h-1}\, a_{kh}D(A_{imkh}) + \sum_{k=i+1}^{n}(-1)^{k-1+h-1}\, a_{kh}D(A_{imkh})\right]$$

$$= -\sum_{i=1}^{n}(-1)^{i+m}\, a_{im}\left[\sum_{k=1}^{i-1}(-1)^{k+h}\, a_{kh}D(A_{imkh}) + \sum_{k=i+1}^{n}(-1)^{k-1+h}\, a_{kh}D(A_{imkh})\right].$$

However if we expand $D(A_{im})$ in terms of its $h$ column we get

$$D(A_{im}) = \sum_{k=1}^{i-1}(-1)^{k+h}\, a_{kh}D(A_{imkh}) + \sum_{k=i+1}^{n}(-1)^{k-1+h}\, a_{kh}D(A_{imkh}).$$

Note that since $m > h$ the elements that make up the $h$ column of $A_{im}$ lie in the $h$ column of $A$ unlike the situation we had in the expansion of $D(B_{ih})$. Therefore

$$\sum_{i=1}^{n}(-1)^{i+j}\, b_{ij}D(B_{ij}) = -\sum_{i=1}^{n}(-1)^{i+m}\, a_{im}D(A_{im}),$$

proving that $D(B) = -D(A)$. The proof that this result holds for $j = m$ and $m > h$ is similar. This completes the proof of the skew symmetry of the mapping for $n \times n$ matrices.

We now turn to the linearity of the mapping. Let $B$ be the matrix obtained from $A$ by multiplying the $k$ column of $A$ by the scalar $c$. Then if $j \neq k$, $b_{ij} = a_{ij}$ and $D(B_{ij}) = cD(A_{ij})$ since the $k$ column of $B_{ij}$ is $c$ times the $k$ column of $A_{ij}$, the other columns being identical. Otherwise $b_{ij} = ca_{ij}$ and $D(B_{ij}) = D(A_{ij})$ since the minors $B_{ij}$ and $A_{ij}$ are identical. Hence

$$\sum_{i=1}^{n}(-1)^{i+j}\, b_{ij}D(B_{ij}) = c\sum_{i=1}^{n}(-1)^{i+j}\, a_{ij}D(A_{ij}).$$

The proof of the remainder of the linearity of the mapping is left as an exercise.

We shall now prove that the mapping satisfies the requirements of the second axiom. Let $A = I_n$, the identity $n \times n$ matrix. Then $A = (\delta_{ij})$. Hence

$$\sum_{i=1}^{n}(-1)^{i+j}\, a_{ij}D(A_{ij}) = \sum_{i=1}^{n}(-1)^{i+j}\delta_{ij}D(A_{ij})$$

$$= (-1)^{2j}\, D(A_{jj}) = D(I_{n-1}) = 1,$$

showing that the second axiom is satisfied by the mapping.

Thus $D$ is the determinant for $n \times n$ matrices. By mathematical induction it is the determinant for all square matrices.

We are now in a position to prove the existence of the determinant.

The previous theorem tells us that determinants of $n \times n$ matrices exist if determinants of $n-1 \times n-1$ matrices exist. We have seen that determinants of $1 \times 1$ matrices exist. By the principle of mathematical induction it follows that the determinant exists.

*Example 2*

Find the determinant of the matrix

$$A = \begin{pmatrix} 0 & 1 & 2 \\ 3 & 4 & 1 \\ 2 & 1 & 0 \end{pmatrix}.$$

We use theorem 1 and have the choice of expanding in terms of any column. Expanding in terms of the first column,

$$\begin{aligned} D(A) &= \sum_{i=1}^{3} (-1)^{i+1} a_{i1} D(A_{i1}) = a_{11}D(A_{11}) - a_{21}D(A_{21}) + a_{31}D(A_{31}) \\ &= 0\begin{vmatrix} 4 & 1 \\ 1 & 0 \end{vmatrix} - 3\begin{vmatrix} 1 & 2 \\ 1 & 0 \end{vmatrix} + 2\begin{vmatrix} 1 & 2 \\ 4 & 1 \end{vmatrix} \\ &= 0 - 3[(1 \times 0) - (1 \times 2)] + 2[(1 \times 1) - (4 \times 2)] \\ &= -8. \end{aligned}$$

To illustrate that the determinant can be expanded using any column let us expand the same determinant from the third column.

$$\begin{aligned} D(A) &= \sum_{i=1}^{3} (-1)^{i+3} a_{i3} D(A_{i3}) = a_{13}D(A_{13}) - A_{23}D(A_{23}) + A_{33}D(A_{33}) \\ &= 2\begin{vmatrix} 3 & 4 \\ 2 & 1 \end{vmatrix} - 1\begin{vmatrix} 0 & 1 \\ 2 & 1 \end{vmatrix} + 0\begin{vmatrix} 0 & 1 \\ 3 & 4 \end{vmatrix} \\ &= 2[(3 \times 1) - (2 \times 4)] - 1[(0 \times 1) - (2 \times 1)] + 0 \\ &= -8. \end{aligned}$$

In practice one expands the determinant from the column containing the most zeros to reduce the work involved.

## Exercises

Evaluate the following determinants.

1 $\begin{vmatrix} 1 & 2 \\ 3 & 4 \end{vmatrix}$.

2 $\begin{vmatrix} -1 & 2 \\ 3 & 1 \end{vmatrix}$.

3 $\begin{vmatrix} 1 & 2 & 3 \\ 0 & 1 & 2 \\ 4 & 1 & 0 \end{vmatrix}$, expanding from the first column.

4 $\begin{vmatrix} a & a^2 & a^3 \\ b & b^2 & b^3 \\ c & c^2 & c^3 \end{vmatrix}$, expanding from the second column.

5 $\begin{vmatrix} a & a & b \\ a & a & c \\ d & d & d \end{vmatrix}$.

6 $\begin{vmatrix} 2 & 1 & 0 \\ 0 & 1 & 2 \\ 3 & 1 & 2 \end{vmatrix}$, expanding from the third column.

7 Show by evaluating the determinants that $\begin{vmatrix} 2 & 1 & 0 \\ 0 & 1 & 2 \\ 3 & 1 & 2 \end{vmatrix} = -\begin{vmatrix} 2 & 0 & 1 \\ 0 & 2 & 1 \\ 3 & 2 & 1 \end{vmatrix}$, illustrating the skew-symmetry property of the determinant in the second and third columns.

8 Show by evaluating the determinants that $\begin{vmatrix} 2 & 1 & 0 \\ 0 & 1 & 2 \\ 3 & 1 & 2 \end{vmatrix} = 1/6 \begin{vmatrix} 2 & 1 & 0 \\ 0 & 1 & 12 \\ 3 & 1 & 12 \end{vmatrix}$, illustrating a linear property of determinants that follows form the first axiom of determinants.

9 Show by evaluating the determinants that $\begin{vmatrix} 2 & 1 & 0 \\ 0 & 1 & 2 \\ 3 & 1 & 2 \end{vmatrix} = \begin{vmatrix} 2 & 4 & 0 \\ 0 & 5 & 2 \\ 3 & 2 & 2 \end{vmatrix} + \begin{vmatrix} 2 & -3 & 0 \\ 0 & -4 & 2 \\ 3 & -1 & 2 \end{vmatrix}$, illustrating the second linear property of determinants that follows from the first axiom.

## 5.3 The Permutation Symbol

In this section we introduce permutations and the permutation symbol. Those properties of permutations that illustrate concepts presented previously in this text and properties that are useful for the discussion of determinants are given.

A *permutation* of the elements of a non-empty, finite set $U$ is defined to be a bijective mapping $\pi$ of $U$ onto $U$. This definition is consistent with one's notion of a permutation as a rearrangement of objects among themselves. The mapping defines the rearrangement.

The set of all permutations on such a set $U$, consisting of $n$ elements, forms a group under the operation of composition. The identity mapping on $U$ is the identity element of this group; it is called the *identity permutation*. Since every element in the set is bijective each has an inverse. The other group properties are easily seen to be true. This group is called the *symmetric* group of degree $n$ and is denoted $S_n$.

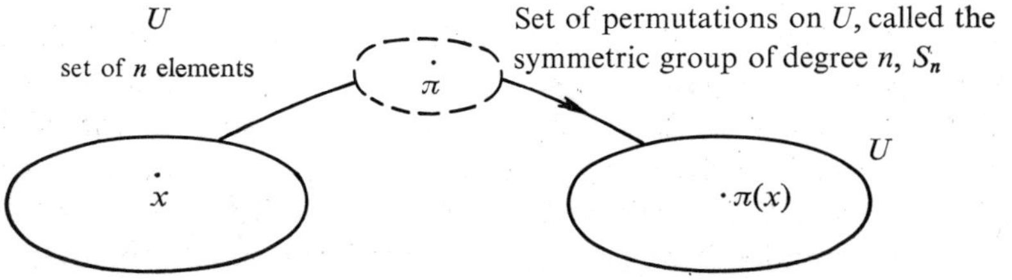

Here we shall be interested in the permutations of the set $\{1, 2, \ldots, n\}$. A permutation $\pi$ of such a set is usually denoted

$$\begin{pmatrix} 1 & 2 & \ldots & n \\ \pi(1) & \pi(2) & \ldots & \pi(n) \end{pmatrix}.$$

For example if $\pi$ is the permutation of the set $\{1, 2, 3, 4\}$ such that $\pi(1) = 3$, $\pi(2) = 2$, $\pi(3) = 4$, $\pi(4) = 1$, $\pi$ is represented thus

$$\begin{pmatrix} 1 & 2 & 3 & 4 \\ 3 & 2 & 4 & 1 \end{pmatrix}.$$

The following example illustrates the product (composite) of two permutations.

*Example 1*

Let $\pi_1$ and $\pi_2$ be the permutations $\begin{pmatrix} 1 & 2 & 3 & 4 \\ 2 & 4 & 3 & 1 \end{pmatrix}$ and $\begin{pmatrix} 1 & 2 & 3 & 4 \\ 1 & 4 & 3 & 2 \end{pmatrix}$. Find $\pi_2\pi_1$.

We have that $\pi_2\pi_1 = \begin{pmatrix} 1 & 2 & 3 & 4 \\ 1 & 4 & 3 & 2 \end{pmatrix} \begin{pmatrix} 1 & 2 & 3 & 4 \\ 2 & 4 & 3 & 1 \end{pmatrix} = \begin{pmatrix} 1 & 2 & 3 & 4 \\ 4 & 2 & 3 & 1 \end{pmatrix}$ since

$$\pi_2\pi_1(1) = \pi_2(2) = 4,$$

$$\pi_2\pi_1(2) = \pi_2(4) = 2,$$

$$\pi_2\pi_1(3) = \pi_2(3) = 3,$$

$$\pi_2\pi_1(4) = \pi_2(1) = 1.$$

Associated with every permutation $\pi$ of the set $\{1, 2, \ldots, n\}$ is the number $+1$ or $-1$, called the *parity* of the permutation. Parity may thus be interpreted as a mapping of $S_n$ onto $\{-1, +1\}$.

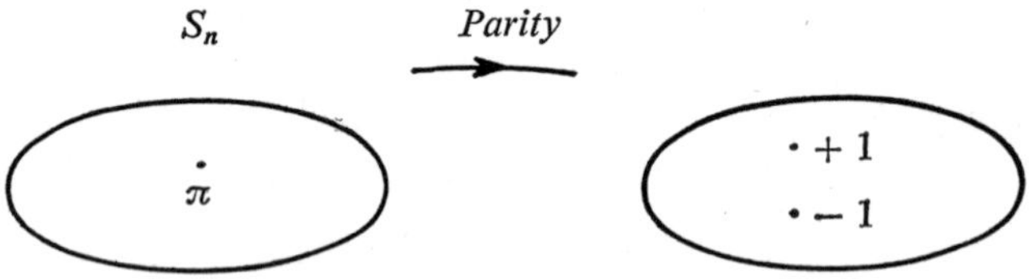


It can be seen that for certain pairs $(p, q)$, $p$ and $q$ elements of $\{1, 2, \ldots, n\}$ with $p < q$, we shall have $\pi(p) < \pi(q)$ while for other pairs with $p < q$ we shall have $\pi(q) < \pi(p)$. This latter case is called an *inversion*. If there are $k$ such inversions the parity of $\pi$ is defined to be $(-1)^k$. If the parity is $+1$ the permutation is said to be *even*, otherwise it is *odd*. Thus to find whether a permutation is even or odd one looks for the number of inversions.

*Example 2*

Determine whether the permutation $\begin{pmatrix} 1 & 2 & 3 & 4 \\ 2 & 3 & 4 & 1 \end{pmatrix}$ is even or odd.

The inversions are the pairs (1, 4), (2, 4) and (2, 3), since $1 < 4$ but $\pi(4) < \pi(1)$, $2 < 4$ but $\pi(4) < \pi(2)$, $3 < 4$ but $\pi(4) < \pi(3)$. For all other pairs $q$ and $p$ with $q < p$ the mapping $\pi$ gives $\pi(q) < \pi(p)$. Hence the number of inversions is 3 and the parity of the permutation is $(-1)^3$, that is $-1$. The permutation is thus an odd permutation.

Let us now define the *permutation symbol.* This is of great theoretical use in the discussion of determinants.

$$\varepsilon_{i_1 i_2 \cdots i_n} \begin{cases} = +1 \quad \text{if} \quad \begin{pmatrix} 1 & 2 & \ldots & n \\ i_1 & i_2 & \ldots & i_n \end{pmatrix} \quad \text{is an even permutation,} \\ = -1 \quad \text{if} \quad \begin{pmatrix} 1 & 2 & \ldots & n \\ i_1 & i_2 & \ldots & i_n \end{pmatrix} \quad \text{is an odd permutation,} \\ = 0 \quad \text{if any of the } i\text{'s are identical.} \end{cases}$$

The permutation symbol is defined only for such indices.

*Example 3*

Find out whether $\varepsilon_{13452} = +1$ or $-1$. We are interested in knowing the parity of the permutation

$$\begin{pmatrix} 1 & 2 & 3 & 4 & 5 \\ 1 & 3 & 4 & 5 & 2 \end{pmatrix}.$$

The inversions of the permutation are (2, 5), (3, 5), (4, 5). Hence the parity is $-1$, and it is an odd permutation.

Therefore $\varepsilon_{13452} = -1$.

*Example 4*

$\varepsilon_{13234} = 0$ since $i_2 = i_4 = 3$.

*Theorem 1*

The permutation symbol is skew-symmetric in its indices. That is, if two indices are interchanged the sign of the symbol is reversed, $\varepsilon_{i_1 \ldots i_p \ldots i_q \ldots i_n} = -\varepsilon_{i_1 \ldots i_q \ldots i_p \ldots i_n}$, for example $\varepsilon_{13452} = -\varepsilon_{13254}$.

Proof: Consider $\varepsilon_{i_1 \ldots i_p \ldots i_q \ldots i_n}$. Assume the indices to be distinct. It depends upon the parity of the permutation $\pi_1$ denoted by

$$\begin{pmatrix} 1 & 2 & \ldots & p & \ldots & q & \ldots & n \\ i_1 & i_2 & \ldots & i_p & \ldots & i_q & \ldots & i_n \end{pmatrix}.$$

$\varepsilon_{i_1 \dots i_q \dots i_p \dots i_n}$ depends upon the parity of the permutation $\pi_2$, denoted by

$$\begin{pmatrix} 1 & 2 & \dots & p & \dots & q & \dots & n \\ i_1 & i_2 & \dots & i_q & \dots & i_p & \dots & i_n \end{pmatrix}.$$

It can be seen that $\pi_1$ and $\pi_2$ differ only in their effect on the elements $p$ and $q$,

$$\pi_1(p) = i_p, \quad \pi_1(q) = i_q,$$

$$\pi_2(p) = i_q, \quad \pi_2(q) = i_p.$$

$\pi_1$ and $\pi_2$ will differ by an odd number of inversions, they have opposite parities. Hence if $\begin{pmatrix} 1\,2 \dots p \dots q \dots n \\ i_1 i_2 \dots i_p \dots i_q \dots i_n \end{pmatrix}$ is an even permutation, then $\begin{pmatrix} 1\,2 \dots p \dots q \dots n \\ i_1 i_2 \dots i_q \dots i_p \dots i_n \end{pmatrix}$ is an odd permutation and vice-versa, proving the theorem for distinct indices.

If two indices are equal then both $\varepsilon_{i_1 \dots i_p \dots i_q \dots i_n}$ and $\varepsilon_{i_1 \dots i_q \dots i_p \dots i_n}$ are identically zero and the theorem holds.

## Exercises

1 Prove that the set of all permutations on $\{1, 2, \dots, n\}$ forms a group under the operation of composition.

2 If $\pi_1 = \begin{pmatrix} 1 & 2 & 3 & 4 \\ 4 & 2 & 3 & 1 \end{pmatrix}$ and $\pi_2 = \begin{pmatrix} 1 & 2 & 3 & 4 \\ 3 & 2 & 1 & 4 \end{pmatrix}$, determine the permutations $\pi_1\pi_2$ and $\pi_2\pi_1$.

3 If $\pi_1 = \begin{pmatrix} 1 & 2 & 3 & 4 & 5 \\ 6 & 3 & 4 & 1 & 2 \end{pmatrix}$ and $\pi_2 = \begin{pmatrix} 1 & 2 & 3 & 4 & 5 \\ 2 & 1 & 3 & 5 & 4 \end{pmatrix}$, determine the permutations $\pi_1\pi_2$ and $\pi_2\pi_1$.

4 Determine whether the following permutations are even or odd.

(a) $\begin{pmatrix} 1 & 2 & 3 \\ 3 & 1 & 2 \end{pmatrix}$, (b) $\begin{pmatrix} 1 & 2 & 3 \\ 1 & 2 & 3 \end{pmatrix}$, (c) $\begin{pmatrix} 1 & 2 & 3 & 4 \\ 4 & 3 & 1 & 2 \end{pmatrix}$, (d) $\begin{pmatrix} 1 & 2 & 3 & 4 & 5 \\ 5 & 4 & 3 & 2 & 1 \end{pmatrix}$.

5 We have proved that the set of all permutations on $\{1, 2, 3, 4, 5\}$ forms a group under the operation of composition. Determine the inverses of the following permutations under this operation.

(a) $\begin{pmatrix} 1 & 2 & 3 & 4 & 5 \\ 4 & 5 & 1 & 3 & 2 \end{pmatrix}$, (b) $\begin{pmatrix} 1 & 2 & 3 & 4 & 5 \\ 3 & 2 & 1 & 5 & 4 \end{pmatrix}$.

6 Prove that the number of odd permutations of $\{1, \dots, n\}$ for $n \geqq 2$ is equal to the number of even permutations.

7 Determine the values of

(a) $\varepsilon_{3412}$, (b) $\varepsilon_{1432}$, (c) $\varepsilon_{53214}$, (d) $\varepsilon_{435421}$.

## 5.4 Properties of Determinants

In this section the permutation symbol is used to develop further properties of determinants. Its use is of theoretical importance in proving that the determinant is a mapping on the set of square matrices having linear properties on the rows of the matrices similar to those defined on the columns. We first of all give an expansion for the determinant of a matrix in terms of the permutation symbol.

*Theorem 1*

If $A$ is the square matrix $(a_{ij})$ then the mapping $D$ defined by $D(A) = \sum_{p=1}^{n} \cdots \sum_{q=1}^{n} \varepsilon_{p \ldots q} a_{1p} \ldots a_{nq}$ is the determinant mapping.

Proof: Because of the third property of the permutation symbol only those terms in which the permutation symbol has distinct indices are non-zero in the above sum.

We shall prove that this mapping satisfies the axioms of the determinant. Hence by the uniqueness of the determinant it must be the determinant. First we show that it is skew-symmetric in the columns of $A$. Let $B = (b_{ij})$ be the matrix obtained from $A$ by interchanging the $h$ and $m$ columns. Then $a_{ij} = b_{ij}$ if $j \neq h, m$; $a_{ih} = b_{im}$ and $a_{im} = b_{ih}$. An arbitrary term in the expansion of $D(A)$ according to the above formula is

$$\begin{aligned} &\varepsilon_{r \ldots h \ldots m \ldots t}\, a_{1r} \ldots a_{kh} \ldots a_{sm} \ldots a_{nt} \\ &= -\varepsilon_{r \ldots m \ldots h \ldots t}\, a_{1r} \ldots a_{kh} \ldots a_{sm} \ldots a_{nt} \\ &= -\varepsilon_{r \ldots m \ldots h \ldots t}\, b_{1r} \ldots b_{km} \ldots b_{sh} \ldots b_{nt}. \end{aligned}$$

This is a term in the expansion of $-D(B)$. In this manner we can associate with every term in the expansion of $D(A)$ a term in the expansion of $-D(B)$. It can also be seen that every term in $-D(B)$ corresponds in this manner to a term in $D(A)$. Hence $D(A) = -D(B)$, proving the required skew-symmetry property.

To prove the first part of the linearity requirement consider the matrix with columns $A^1, \ldots, A^i + A'^i, \ldots, A^n$. An arbitrary element in the expansion of $D(A^1, \ldots, A^i + A'^i, \ldots, A^n)$ may be written

$$\begin{aligned} &\varepsilon_{r \ldots i \ldots t} a_{1r} \ldots (a_{ki} + a'_{ki}) \ldots a_{nt} \\ &= \varepsilon_{r \ldots i \ldots t} a_{1r} \ldots a_{ki} \ldots a_{nt} + \varepsilon_{r \ldots i \ldots t} a_{1r} \ldots a'_{ki} \ldots a_{nt}. \end{aligned}$$

This expression is the sum of an element in the expansion of $D(A^1, \ldots, A^i, \ldots, A^n)$ and an element of $D(A^1, \ldots, A'^i, \ldots, A^n)$. Furthermore, it can be seen that the terms in the expansion of $D(A^1, \ldots, A^i, \ldots, A^n)$ and $D(A^1, \ldots, A'^i, \ldots, A^n)$ can be grouped together in this manner so that an arbitrary element of $D(A^1, \ldots, A^i, \ldots, A^n) + D(A^1, \ldots, A'^i, \ldots, A^n)$ corresponds to an element of $D(A^1, \ldots, A^i + A'^i, \ldots, A^n)$. Hence $D(A^1, \ldots, A^i + A'^i, \ldots, A^n) = D(A^1, \ldots, A^i, \ldots, A^n) + D(A^1, \ldots, A'^i, \ldots, A^n)$, and the mapping $D$ defined here satisfies this linearity requirement of the determinant.

Further, for the matrix with columns $A^1, \ldots, cA^i, \ldots, A^n$, where $c$ is a scalar, an aıbitrary term in the expansion of $D(A^1, \ldots, cA^i, \ldots, A^n)$ is

$$\varepsilon_{r\ldots i\ldots t} a_{1r} \ldots (ca_{ki}) \ldots a_{nt}$$
$$= c\,\varepsilon_{r\ldots i\ldots t} a_{1r} \ldots a_{ki} \ldots a_{nt}.$$

This is an element in the expansion of $cD(A^1, \ldots, A^i, \ldots, A^n)$. Hence every term in the expansion of $D(A^1, \ldots, cA^i, \ldots, A^n)$ is a term in the expansion of $cD(A^1, \ldots, A^i, \ldots, A^n)$. Furthermore it can be seen that each term in $cD(A^1, \ldots, A^i, \ldots, A^n)$ corresponds in this manner to a term in

$$D(A^1, \ldots, cA^i, \ldots, A^n).$$

Hence $D(A^1, \ldots, cA^i, \ldots, A^n) = cD(A^1, \ldots, A^i, \ldots, A^n)$ and the mapping $D$ is linear in the columns of the matrix.

Finally, if $A$ is the unit matrix there is only one non-zero term in the expansion of $D(A)$, namely the term $\varepsilon_{1\ldots n} a_{11} \ldots a_{nn}$. This term is 1, proving that the mapping $D$ gives $D(I_n) = 1$.

Hence $D$ as defined is the determinant mapping.

*Theorem 2*

The determinant mapping is a skew-symmetric multilinear mapping on the rows of a square matrix into the field of scalars.

Proof: Let us use subscripts to denote rows of a matrix. If $A$ is the matrix with rows $A_1, \ldots, A_i, \ldots, A_j, \ldots, A_n$,

$$\begin{aligned} D(A) = D(A_1, \ldots, A_i, \ldots, A_j, \ldots, A_n) &= \sum_{p=1}^{n} \cdots \sum_{r=1}^{n} \cdots \sum_{s=1}^{n} \cdots \sum_{q=1}^{n} \varepsilon_{p\ldots r\ldots s\ldots q} \\ &\qquad a_{1p} \ldots a_{ir} \ldots a_{js} \ldots a_{nq} \\ &= -\sum_{p=1}^{n} \cdots \sum_{s=1}^{n} \cdots \sum_{r=1}^{n} \cdots \sum_{q=1}^{n} \varepsilon_{p\ldots s\ldots r\ldots q} \\ &\qquad a_{1p} \ldots a_{js} \ldots a_{ir} \ldots a_{nq} \\ &= -D(A_1, \ldots, A_j, \ldots, A_i, \ldots, A_n), \end{aligned}$$

proving that the determinant is skew-symmetric in the rows of $A$.

Consider a matrix with rows $A_1, \ldots, A_i + B_i, \ldots, A_n$. Then

$$\begin{aligned} D(A_1, \ldots, A_i + B_i, \ldots, A_n) &= \sum_{p=1}^{n} \cdots \sum_{r=1}^{n} \cdots \sum_{q=1}^{n} \varepsilon_{p\ldots r\ldots q} a_{1p} \ldots (a_{ir} + b_{ir}) \ldots a_{nq} \\ &= \sum_{p=1}^{n} \cdots \sum_{r=1}^{n} \cdots \sum_{q=1}^{n} \varepsilon_{p\ldots r\ldots q} a_{1p} \ldots a_{ir} \ldots a_{nq} \\ &\quad + \sum_{p=1}^{n} \cdots \sum_{r=1}^{n} \cdots \sum_{q=1}^{n} \varepsilon_{p\ldots r\ldots q} a_{1p} \ldots b_{ir} \ldots a_{nq} \\ &= D(A_1, \ldots, A_i, \ldots, A_n) + D(A_1, \ldots, B_i, \ldots A_n) \end{aligned}$$

Further, for the matrix with rows $A_1, \ldots, cA_i, \ldots, A_n$ where $c$ is a scalar,

$$D(A_1, \ldots, cA_i, \ldots, A_n) = \sum_{p=1}^{n} \cdots \sum_{r=1}^{n} \cdots \sum_{q=1}^{n} \varepsilon_{p\ldots r\ldots q} a_{1p} \cdots (ca_{ir}) \cdots a_{nq}$$

$$= c\sum_{p=1}^{n} \cdots \sum_{r=1}^{n} \cdots \sum_{q=1}^{n} \varepsilon_{p\ldots r\ldots q} a_{1p} \cdots a_{ir} \cdots a_{nq}$$

$$= cD(A_1, \ldots, A_i, \ldots, A_n).$$

This proves the linearity of the determinant in the rows of $A$.

It can further be proved that the determinant can alternatively be defined to be a skew-symmetric multilinear mapping on the rows of a square matrix which maps each unit matrix into the multiplicative identity of the field of scalars.

*Example 1*

On evaluating the determinants it can be seen that

$$\begin{vmatrix} 0 & 1 & 2 \\ 3 & 4 & 1 \\ 2 & 4 & 1 \end{vmatrix} = 7 \quad \text{and} \quad \begin{vmatrix} 0 & 1 & 2 \\ 2 & 4 & 1 \\ 3 & 4 & 1 \end{vmatrix} = -7,$$

illustrating the skew symmetry of the determinant in the second and third rows.

*Example 2*

$$\begin{vmatrix} 0 & 1 & 2 \\ 3 & 4 & 1 \\ 2 & 4 & 1 \end{vmatrix} = 7 \quad \text{and} \quad \begin{vmatrix} 0 & 1 & 2 \\ 3 & 4 & 1 \\ 4 & 8 & 2 \end{vmatrix} = 14$$

giving $\begin{vmatrix} 0 & 1 & 2 \\ 3 & 4 & 1 \\ 2 & 4 & 1 \end{vmatrix} = \frac{1}{2}\begin{vmatrix} 0 & 1 & 2 \\ 3 & 4 & 1 \\ 4 & 8 & 2 \end{vmatrix}$, illustrating the linearity of the determinant in the third row.

*Theorem 3*

The determinant of a matrix can be expanded in terms of the $i$th row according to the following rule,

$$D(A) = \sum_{j=1}^{n} (-1)^{i+j} a_{ij} D(A_{ij}).$$

In the theorem of section 5.2 we had a rule for expanding the determinant of a matrix in terms of any column. This theorem implies that there is an analogous rule for expanding in terms of any row.

The proof is similar to that of the theorem in section 5.2 with the roles of columns and rows interchanged, interpreting the determinant as a skew-symmetric mapping on the rows of the matrix.

*Example 3*

Expand the determinant of the matrix

$$\begin{pmatrix} 1 & 2 & 3 \\ 0 & 1 & 2 \\ 1 & 2 & 0 \end{pmatrix}$$

in terms of the third row.

$$\begin{aligned} D(A) &= \sum_{j=1}^{3} (-1)^{3+j} a_{3j} D(A_{3j}) \\ &= a_{31}D(A_{31}) - a_{32}D(A_{32}) + a_{33}D(A_{33}) \\ &= 1\begin{vmatrix} 2 & 3 \\ 1 & 2 \end{vmatrix} - 2\begin{vmatrix} 1 & 3 \\ 0 & 2 \end{vmatrix} + 0\begin{vmatrix} 1 & 2 \\ 0 & 1 \end{vmatrix} \\ &= 1[(2 \times 2) - (1 \times 3)] - 2[(1 \times 2) - (0 \times 3)] + 0 \\ &= -3. \end{aligned}$$

*Theorem 4*

If $A$ and $B$ are square matrices with $B$ being the transpose of $A$, then $D(A) = D(B)$.

Proof: We shall proceed by induction. The theorem is true for $1 \times 1$ matrices.

Assume the theorem is true for all square matrices of order $n - 1$. We shall show that it then holds for matrices of order $n$ and hence by induction it holds for all square matrices.

Let $A$ and $B$ be $n \times n$ matrices. Expand $D(A)$ in terms of the first row to get

$$D(A) = \sum_{j=1}^{n} (-1)^{1+j} a_{1j} D(A_{1j}). \tag{1}$$

Expand $D(B)$ in terms of the first column to get

$$D(B) = \sum_{i=1}^{n} (-1)^{i+1} b_{i1} D(B_{i1}). \tag{2}$$

But since $B = A^t$, the first row of $A$ is identical to the first column of $B$, that is $a_{1j} = b_{j1}$ for $j = 1, \ldots, n$. $D(A_{1j})$ and $D(B_{j1})$ are both determinants of $(n - 1) \times (n - 1)$ matrices. Since $B = A^t$ the one $(n - 1) \times (n - 1)$ matrix is the transpose of the other. Hence by the hypothesis involving the determinants of square matrices of order $n - 1$, $D(A_{1j}) = D(B_{j1})$ for all $j = 1, \ldots, n$.

The two right sides of identities (1) and (2) are therefore equal giving $D(A) = D(B)$. By induction the theorem holds for all square matrices.

*Example 4*

If $A = \begin{pmatrix} 1 & 2 & 3 \\ 0 & 1 & 2 \\ 3 & 1 & 2 \end{pmatrix}$ then $A^t = \begin{pmatrix} 1 & 0 & 3 \\ 2 & 1 & 1 \\ 3 & 2 & 2 \end{pmatrix}$.

On expanding both determinants it can be seen that $D(A) = D(A^t) = 3$.

We now give a result that can be used to simplify determinants before expanding to evaluate.

*Theorem 5*

If $B$ is obtained from the square matrix $A$ by multiplying the $k$th row (column) by a non-zero scalar $c$ and adding the result to the $i$th row (column) using vector addition, where $i \neq k$, then $D(B) = D(A)$.

Proof: The element $b_{ij}$ of the $i$th row of $B$ is of the form $a_{ij} + ca_{kj}$. All elements in rows other than the $i$th row of $B$ are identical to the corresponding elements in $A$. Hence by the multilinearity of the determinant in the rows of a matrix,

$$D(B) = D(A) + cD(A'),$$

where $A'$ is obtained from $A$ by replacing the $i$th row by the $k$th row. But then the $i$th row and $k$th row of $A'$ are identical giving $D(A') = 0$. Hence $D(B) = D(A)$.

The proof for columns is similar, using the multilinearity of the determinant in the columns of a matrix.

*Example 5*

Evaluate the following determinant:

$$D(A) = \begin{vmatrix} 0 & 1 & 3 & 0 \\ 4 & 5 & 7 & 2 \\ -2 & -1 & -3 & 0 \\ 4 & 2 & 5 & 0 \end{vmatrix}.$$

Multiply the third row by 2 and add to the 4th row to get

$$D(A) = \begin{vmatrix} 0 & 1 & 3 & 0 \\ 4 & 5 & 7 & 2 \\ -2 & -1 & -3 & 0 \\ 0 & 0 & -1 & 0 \end{vmatrix}.$$

Now expand in terms of the last row to get

$$-(-1)\begin{vmatrix} 0 & 1 & 0 \\ 4 & 5 & 2 \\ -2 & -1 & 0 \end{vmatrix} = 1 \begin{vmatrix} 4 & 2 \\ -2 & 0 \end{vmatrix} = 4.$$

In this manner theorem 5 can be used to create zeros in certain slots of the determinant that make its evaluation easier.

This technique is useful for the evaluation of small-order determinants. Computer programs have been written for evaluating large-order determinants.

*Theorem 6*

Consider the operation of multiplication on the set of square matrices of the same kind over a field and the operation of multiplication associated with the field of scalars. The determinant is a homomorphism with respect to these operations.

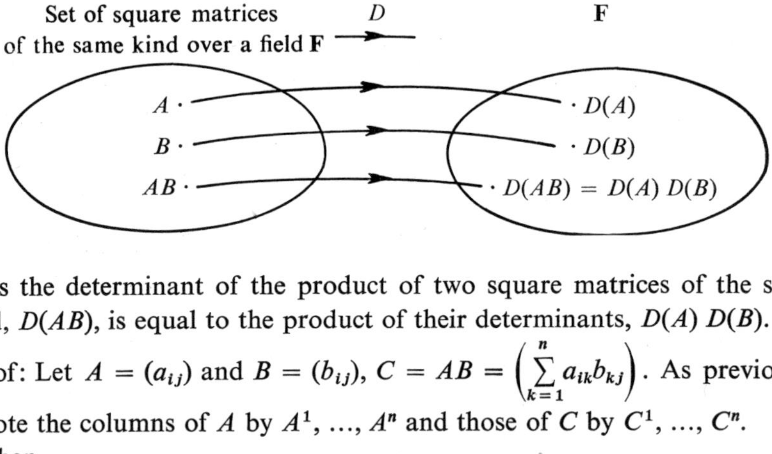


Thus the determinant of the product of two square matrices of the same kind, $D(AB)$, is equal to the product of their determinants, $D(A)\,D(B)$.

Proof: Let $A = (a_{ij})$ and $B = (b_{ij})$, $C = AB = \left(\sum_{k=1}^{n} a_{ik}b_{kj}\right)$. As previously denote the columns of $A$ by $A^1, \ldots, A^n$ and those of $C$ by $C^1, \ldots, C^n$.

Then
$$D(AB) = D(C) = D(C^1, \ldots, C^n)$$
$$= D(b_{11}A^1 + \cdots + b_{n1}A^n, \ldots, b_{1n}A^1 + \cdots + b_{nn}A^n).$$

Using the skew-symmetry and multilinearity of $D$ this can be expressed

$$\sum_{p=1}^{n} \cdots \sum_{q=1}^{n} \varepsilon_{p\cdots q} b_{1p} \cdots b_{nq} D(A^1, \ldots, A^n),$$

which is equal to $D(B)\,D(A)$.

**Exercises**

1 Expand the determinant of the matrix $\begin{pmatrix} 1 & 2 & 0 \\ 3 & 1 & 2 \\ 0 & 1 & 1 \end{pmatrix}$ in terms of the second row and then in terms of the third row to illustrate theorem 3. - 3(2) - 1(1) + 2(1) = 7
0(4) - 1(2) + 1(-5) = -7

2 Use the result of theorem 5 to simplify the following determinants and then evaluate:

(a) $\begin{vmatrix} 1 & 2 & 3 \\ 2 & 4 & 2 \\ 0 & 1 & 1 \end{vmatrix}$ (b) $\begin{vmatrix} 1 & 0 & 3 \\ 1 & 2 & 0 \\ 2 & 3 & 6 \end{vmatrix}$ (c) $\begin{vmatrix} 3 & 1 & 4 & 2 \\ 4 & 3 & 1 & 6 \\ 2 & 2 & 0 & 2 \\ 1 & 0 & 2 & 0 \end{vmatrix}$.

3 Prove that if all the elements of one row (or column) of a square matrix are zero, then the value of the determinant of that matrix is zero.

4 Prove that if two rows (two columns) of a square matrix are proportional then the determinant of that matrix is zero.

5 If $A$ is a non-singular matrix such that $A^2 = A$ determine the value of $D(A)$.

## 5.5 Rank of a Matrix

Let $A$ be an $m \times n$ matrix. The columns of $A$ may be viewed as the set of column vectors $\{A^1, \ldots, A^n\}$ and the rows as the set of row vectors $\{A_1, \ldots, A_m\}$. The *column rank* of $A$ is the number of linearly independent column vectors and the *row rank* is defined analogously.

*Theorem 1*

The row rank and the column rank of a matrix are identical. This integer is called the *rank of the matrix*.

Proof: Let the column rank be $q$ and the row rank $r$. Relabel the columns and rows if necessary so that $A^1, \ldots, A^q$ are $q$ linearly independent columns and $A_1, \ldots, A_r$ are $r$ linearly independent rows. Then each column vector can be expressed as a linear combination of $A^1, \ldots, A^q$. Let

$A^i = \sum_{j=1}^{q} x_{ji} A^j$ for the $i$th column vector.

This is a vector equation that can be expanded and written

$$\begin{pmatrix} a_{1i} \\ a_{2i} \\ \vdots \\ a_{mi} \end{pmatrix} = x_{1i} \begin{pmatrix} a_{11} \\ a_{21} \\ \vdots \\ a_{m1} \end{pmatrix} + \cdots x_{qi} \begin{pmatrix} a_{1q} \\ a_{2q} \\ \vdots \\ a_{mq} \end{pmatrix}.$$

The $k$th equation here is

$$a_{ki} = \sum_{j=1}^{q} x_{ji} a_{kj}.$$

But $a_{ki}$ can be looked upon as the $i$th element in the $k$th row vector $A_k$ and $x_{ji}$ can be looked upon as the $i$th element in a row vector $X_j$.

Hence $A_k = \sum_{j=1}^{q} a_{kj} X_j$.

This means that the row vectors can be expressed as a linear combination of the $q$ vectors $X_1, \ldots, X_q$. Hence the row rank $\leqq$ column rank. It can be proved in a similar manner that the column rank $\leqq$ row rank, implying that the column rank and row rank are identical.

We know that an $n \times n$ matrix with elements from a field $\mathbf{F}$ can be interpreted as a linear mapping of $\mathbf{F}^n$ into itself. We now relate the rank of a matrix to its properties as a mapping.

*Theorem 2*

Let $A$ be an $n \times n$ matrix over a field $\mathbf{F}$. Then, rank of $A$ is $n \Leftrightarrow A$, interpreted as a mapping of $\mathbf{F}^n$ into itself, is an isomorphism.

Proof: Let $A$ be a matrix of rank $n$, interpreted as a mapping of $\mathbf{F}^n$ into $\mathbf{F}^n$. We already know that $A$ mapping $\mathbf{F}^n$ into $\mathbf{F}^n$ with operation of $A$ on $\mathbf{F}^n$ defined by the usual pre-multiplication of a vector by a matrix is a linear mapping (example 4, section 3.2). It remains to prove that this mapping is bijective.

If $v$ is an element of $\mathbf{F}^n$ then there exist scalars $a_1, \ldots, a_n$, such that $v = a_1A^1 + \cdots + a_nA^n$, since the rank of $A$ is $n$ and its columns $A^1, \ldots, A^n$ thus form a basis for $\mathbf{F}^n$. This implies that $v = A\begin{pmatrix} a_1 \\ \vdots \\ a_n \end{pmatrix}$ and hence that $A$ is surjective, mapping the element $\begin{pmatrix} a_1 \\ \vdots \\ a_n \end{pmatrix}$ of $\mathbf{F}^n$ into $v$.

To prove that $A$ is injective consider distinct elements $u_1$ and $u_2$ of $\mathbf{F}^n$. Let $u_1 = \begin{pmatrix} b_1 \\ \vdots \\ b_n \end{pmatrix}$ and $u_2 = \begin{pmatrix} c_1 \\ \vdots \\ c_n \end{pmatrix}$. Then

$$Au_1 = b_1A^1 + \cdots + b_nA^n,$$

and

$$Au_2 = c_1A^1 + \cdots + c_nA^n.$$

$$\begin{aligned} Au_1 - Au_2 &= b_1A^1 + \cdots + b_nA^n - c_1A^1 \ldots - c_nA^n \\ &= (b_1 - c_1)\,A^1 + \cdots + (b_n - c_n)\,A^n. \end{aligned}$$

Since $u_1 \neq u_2$, $b_i \neq c_i$ for some $1 \leqq i \leqq n$. Therefore, since $A^1, \ldots, A^n$ are linearly independent

$$(b_1 - c_1)\,A^1 + \cdots + (b_n - c_n)\,A^n \neq O,$$

proving that $Au_1 - Au_2 \neq O$. Hence $Au_1 \neq Au_2$, and $A$ is injective.

Thus $A$ is an isomorphism of $\mathbf{F}^n$ onto $\mathbf{F}^n$.

To see that the converse holds we use the injective property of $A$. If $u_1$ and $u_2$ are distinct elements, as above, then since $A$ is injective, $Au_1 \neq Au_2$.

This implies that

$$A(u_1 - u_2) \neq O,$$

giving that

$$(b_1 - c_1)\, A^1 + \cdots + (b_n - c_n)\, A^n \neq O.$$

Since the vectors $u_1$ and $u_2$ are arbitrary distinct vectors, the coefficients $(b_1 - c_1), \ldots, (b_n - c_n)$ are arbitrary scalars, not all zero. Hence the column vectors of $A$, $A^1, \ldots, A^n$ are linearly independent, proving that $A$ is of rank $n$.

*Theorem 3*

Let $A$ be an $n \times n$ matrix over a field $\mathbf{F}$. The rank of $A$ is $n \Leftrightarrow A$ is non-singular.

Proof: Let $A$ be of rank $n$. It is thus an isomorphism and has an inverse $A^{-1}$ satisfying $AA^{-1} = I_n$. (We prove the existence of this inverse matrix in the next chapter, section 6.1). Taking determinants, $D(AA^{-1}) = 1$, giving $D(A)\, D(A^{-1}) = 1$. This implies that $D(A) \neq o$.

To prove the converse assume that the rank of $A$ is less than $n$. Then one row is linearly dependent upon the other rows. By repeated application of theorem 5, section 5.4, one can therefore make one row of the determinant of $A$ identically zero, giving the determinant of $A$ to be zero. This contradicts $A$ being non-singular. Hence the assumption was incorrect and the rank of $A$ is $n$.

Theorems 2 and 3 can be combined to give: rank of $A$ is $n \Leftrightarrow A$ is an isomorphism interpreted as a mapping of $\mathbf{F}^n$ into itself $\Leftrightarrow A$ is non-singular.

## Exercises

1 Determine the ranks of the following matrices:

(a) $\begin{pmatrix} 1 & 2 \\ 3 & 4 \end{pmatrix}$, (b) $\begin{pmatrix} 1 & 2 & 1 \\ 2 & 4 & 2 \\ 1 & 2 & 3 \end{pmatrix}$, (c) $\begin{pmatrix} 1 & 0 & 0 \\ 0 & 1 & 0 \\ 0 & 0 & 1 \end{pmatrix}$, (d) $\begin{pmatrix} 2 & 1 & 3 \\ 4 & 2 & 6 \\ 2 & 1 & 3 \end{pmatrix}$.

2 Discuss the interpretation of rank as a mapping.

3 If $A$ and $B$ are matrices of the same kind prove that

$$\text{rank}\,(A + B) \leq \text{rank}\, A + \text{rank}\, B.$$

4 An $m \times n$ matrix $A$ over a field $\mathbf{F}$ can be interpreted as a linear mapping of $\mathbf{F}^n$ into $\mathbf{F}^m$, where the operation of $A$ on an element $\begin{pmatrix} x_1 \\ \vdots \\ x_n \end{pmatrix}$ of $\mathbf{F}^n$ is defined by $A \begin{pmatrix} x_1 \\ \vdots \\ x_n \end{pmatrix}$, using matrix multiplication. Prove that the rank of $A$ is equal to the dimension of the image space of this mapping.

## 5.6 Existence and Uniqueness of Solutions to Systems of Linear Equations[1]

Consider the following system of $m$ linear equations in $n$ unknowns:

$$\begin{aligned} a_{11}x_1 + a_{12}x_2 + \cdots + a_{1n}x_n &= y_1 \\ \vdots \qquad\qquad &\quad \vdots \\ a_{m1}x_1 + a_{m2}x_2 + \cdots + a_{mn}x_n &= y_m. \end{aligned}$$

Let $A$ be the matrix of coefficients, $X$ the column vector $\begin{pmatrix} x_1 \\ \vdots \\ x_n \end{pmatrix}$ and $Y$ the vector $\begin{pmatrix} y_1 \\ \vdots \\ y_n \end{pmatrix}$. The system can be written as one matrix equation

$$AX = Y.$$

Hence interpreting $A$ as a linear mapping of $\mathbf{F}^n$ into itself, the set of solutions is the subset of $\mathbf{F}^n$ which is mapped linearly onto the vector $Y$.

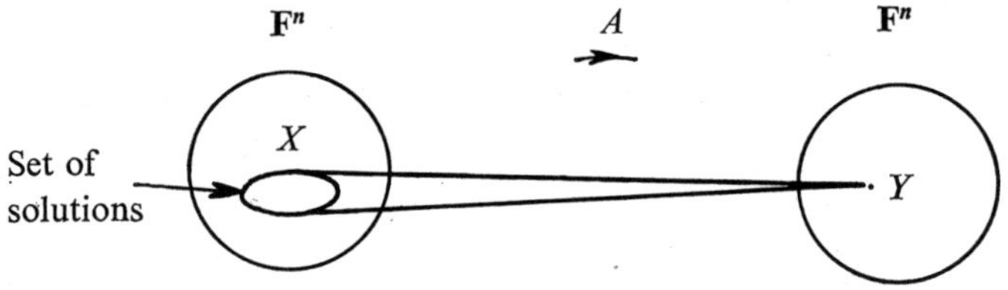


In the above system the matrix

$$\begin{pmatrix} a_{11} & \cdots & a_{1n} y_1 \\ \vdots & & \\ a_{m1} & \cdots & a_{mn} y_m \end{pmatrix}$$

is called the *augmented matrix*.

We are now in a position to discuss the existence and uniqueness of solutions to such systems.

*Theorem 1*

A system of $n$ linear equations in $n$ unknowns has a solution if and only if the rank $r$ of the augmented matrix is equal to that of the matrix of coefficients.

---

[1] The theory of systems of linear equations is a special case within the theory of linear inequality systems. Such systems are discussed in *linear programming*. Linear programming is applied extensively to a broad class of decision problems encountered in government, industry, economies and engineering. The reader is referred to George B. Dantzig, *Linear Programming and Extensions*. (Princeton University Press, 1963). This book gives a discussion of linear programming that includes its historical development.

Proof: Let $\{A^1, \ldots, A^n\}$ be the column vectors of the matrix of coefficients. Then the system of equations can be written

$$x_1A^1 + \cdots + x_nA^n = Y.$$

It can be seen that a solution exists if and only if $Y$ is linearly dependent on $A^1, \ldots, A^n$, that is if the rank of the augmented matrix is the same as that of the matrix of coefficients.

*Theorem 2*

a) If $Y = 0$ the system of equations is said to be *homogeneous* and the set of solutions is the kernel of $A$.

b) For $Y \neq O$ the system of equations is *non-homogeneous.* If $X$ is a solution of $AX = Y$ then every solution can be written $X + Z$ where $Z$ is an element of the kernel of $A$ (see exercise 4, section 3.4).

Proof: In the first part of this theorem the system may be written in the form $AX = O$. It follows that the set of solutions is the kernel of $A$.

For the second part of the theorem let $X_1$ be an arbitrary solution to the system of equations. Hence $AX_1 = Y$. Further $X$ is a given solution, therefore $AX = Y$. These two equations can be combined to give $AX_1 = AX$. By the linearity of $A$ this implies that $A(X_1 - X) = O$, giving $X_1 - X$ to be an element $Z$ of the kernel of $A$. An arbitrary solution $X_1$ can therefore be written $X + Z$, proving the theorem.

*Theorem 3*

If the matrix of coefficients and the augmented matrix of a system of $n$ linear equations in $n$ unknowns have the same rank $r$ and $r = n$, then the solution is unique.

Proof: A solution exists since the matrix of coefficients and augmented matrix have the same rank. Further, since the rank of the matrix of coefficients is $n$ the column vectors $A^1, \ldots, A^n$ are linearly independent and $Y$ can therefore be expressed uniquely as a linear combination of them, $x_1A^1 + \cdots + x_nA^n = Y$.

*Theorem 4*

If the matrix of coefficients and the augmented matrix of a system of $n$ linear equations in $n$ unknowns have the same rank $r$ and $r < n$ then the solution is not unique.

Proof: It was seen in the proof of theorem 2 of the previous section that if a matrix is injective then its rank is $n$. Here the rank of the matrix of coefficients, $A$, is less than $n$, hence it is not injective. The kernel of $A$ is thus not the zero vector (theorem 4, section 3.4). Hence the solution is not unique: any element of the form $X + Z$ will be a solution, where $X$ is a particular solution and $Z$ is an element of the kernel.

### Exercises

Discuss the existence and uniqueness of solutions to the following systems of linear equations:

1
$$\begin{aligned} 3x_1 + 2x_2 + 4x_3 &= 1 \\ x_1 + \phantom{2}x_3 &= 0 \\ 2x_1 + x_2 + 2x_3 &= 2 \end{aligned}$$

2
$$\begin{aligned} 2x_1 + 3x_2 &= 4 \\ x_1 - 2x_2 &= 1 \\ 3x_1 + 5x_2 &= 2 \end{aligned}$$

3
$$\begin{aligned} x_1 + 2x_2 + 4x_3 &= 1 \\ 2x_1 + 3x_2 + 2x_3 &= 2 \end{aligned}$$

4
$$\begin{aligned} 2x_1 + 3x_2 + 2x_3 &= 2 \\ x_1 + 2x_2 + 3x_3 &= 0 \\ 5x_1 + 3x_2 + x_3 &= 0 \end{aligned}$$

5
$$\begin{aligned} x_1 + 2x_2 + 3x_3 &= 0 \\ 2x_1 + 4x_2 + 6x_3 &= 0 \\ -x_1 - 2x_2 - 3x_3 &= 0 \end{aligned}$$

## 5.7 Solutions to Systems of Linear Equations

In this section we give a method for finding solutions to systems of linear equations using transformations of matrices.

With every system of $m$ equations in $n$ variables we can associate the $m \times (n+1)$ augmented matrix. Conversely every $m \times (n+1)$ matrix can be interpreted as the augmented matrix of a system of $m$ equations in $n$ variables. This defines a bijective mapping $b$ of the set consisting of all such systems onto the set of $m \times (n+1)$ matrices.

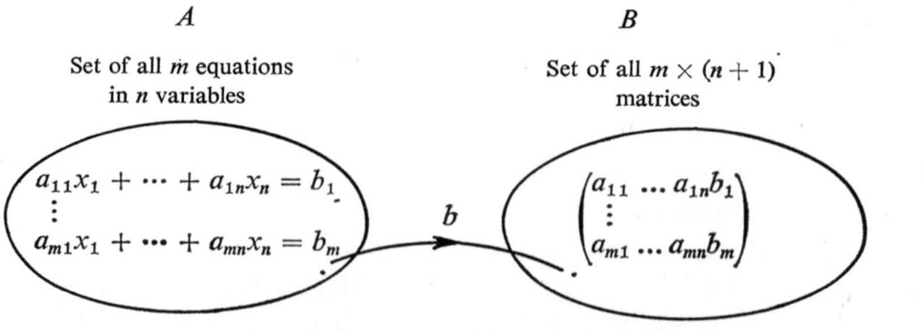


One method of solving the system of equations is to manipulate the equations in certain ways, eliminating variables. Here we axiomize these manipulations and define analogous transformations on the set of matrices.

We use these transformations to arrive at a simpler system of equations that has the same solutions as the original system. (We shall include the possibility of only one solution when we talk about solutions.)

The manipulations that are used for solving systems of linear equations are

1 Interchanging two equations.
2 Multiplying an equation of the system by a non-zero constant.
3 Multiplying an equation by a constant and adding it to another equation.

Such manipulations are called *elementary transformations*. They lead to new systems which have the same solutions as the original one. If the original system has no solutions then the new system has no solutions and conversely. Each elementary transformation is a mapping of an element of the set $A$ of $m$ equations in $n$ variables onto an element of $A$. Systems of linear equations having the same solutions are called *equivalent systems*. Systems consisting of the same number of equations in the same number of variables and having no solutions are also called equivalent systems. (Readers who are not familiar with these results should do exercises 10 and 11 of this section.)

*Theorem 1*

The composite of two elementary transformations leads to a system which is equivalent to the original one.
Proof: Let $T_1$ be an elementary transformation that maps the system $S_1$ onto $S_2$ and $T_2$ an elementary transformation that maps $S_2$ onto $S_3$. Then $S_1$ and $S_2$ have the same (or no) solutions and so do $S_2$ and $S_3$. Hence $T_2 \circ T_1$ maps $S_1$ onto $S_3$, systems that have the same (or no) solutions.

We now develop a theory of analogous transformations for the augmented matrices associated with the systems of linear equations. Let $T$ be an elementary transformation that maps a system $S_1$ onto $S_2$. If $M_1$ and $M_2$ are the augmented matrices, define a mapping $t$ of $M_1$ onto $M_2$. The following diagram illustrates the concept.

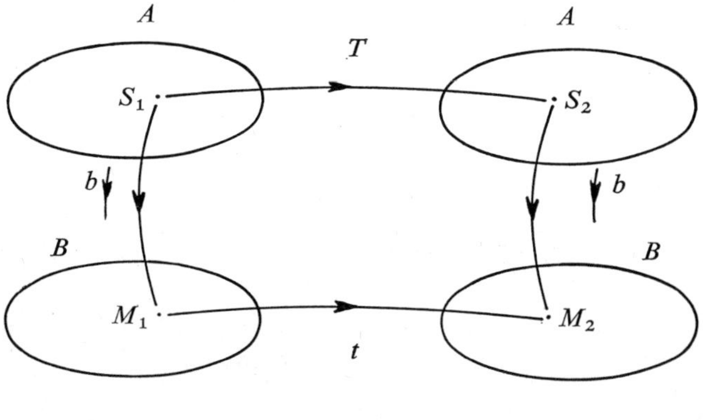

Call such a mapping $t$ an *elementary matrix transformation* on the matrix $M_1$. $M_1$ and $M_2$ are said to be *equivalent matrices* and we write $M_1 \cong M_2$. It can be seen that elementary matrix transformations are transformations of the following kinds:

1 Interchanging two rows of the matrix.

2 Multiplying a row by a non-zero constant.

3 Multiplying a row by a constant and adding it to another row.

The following theorem is a consequence of these definitions and of theorem 1.

*Theorem 2*

Let $t_1$ be an elementary matrix transformation that maps $M_1$ onto $M_2$, $t_2$ an elementary matrix transformation that maps $M_2$ onto $M_3$. Then $t_2 \circ t_1$ maps $M_1$ onto $M_3$ and the systems of linear equations associated with the matrices $M_1$ and $M_3$ are equivalent.

The proof of this theorem is given as an exercise.

The use of this theory is in reducing the system of equations to a simpler system by means of the transformations on matrices. One is able to create zeros in certain positions for the matrices by means of elementary transformations and then consider an equivalent simpler system of equations. The following figure illustrates the method.

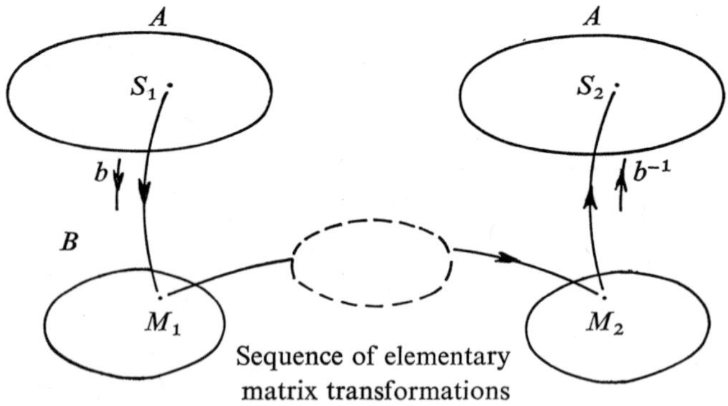


Given the system $S_1$ of linear equations one transforms the augmented matrix $M_1$ using a sequence of elementary matrix transformations into a matrix $M_2$ then solves a simpler system of equivalent equations $S_2$. This method is called *the method of Gaussian elimination*.

*Example 1*

Solve if possible the system of equations

$$\begin{aligned} x_1 + x_2 + x_3 &= 3 \\ x_1 - x_2 - 2x_3 &= -5 \\ 2x_1 + 3x_2 + x_3 &= 5. \end{aligned}$$

The augmented matrix is

$$\begin{pmatrix} 1 & 1 & 1 & 3 \\ 1 & -1 & -2 & -5 \\ 2 & 3 & 1 & 5 \end{pmatrix}.$$

We have, using elementary transformations,

$$\begin{pmatrix} 1 & 1 & 1 & 3 \\ 1 & -1 & -2 & -5 \\ 2 & 3 & 1 & 5 \end{pmatrix} \underset{\text{Row } 2 - \text{Row } 1}{\cong} \begin{pmatrix} 1 & 1 & 1 & 3 \\ 0 & -2 & -3 & -8 \\ 2 & 3 & 1 & 5 \end{pmatrix} \underset{\text{Row } 3 - 2\,(\text{Row } 1)}{\cong} \begin{pmatrix} 1 & 1 & 1 & 3 \\ 0 & -2 & -3 & -8 \\ 0 & 1 & -1 & -1 \end{pmatrix}$$

$$\underset{\text{Row } 3 + 1/2(\text{Row } 2)}{\cong} \begin{pmatrix} 1 & 1 & 1 & 3 \\ 0 & -2 & -3 & -8 \\ 0 & 0 & -\frac{5}{2} & -5 \end{pmatrix} \underset{-2/5\ (\text{Row } 3)}{\cong} \begin{pmatrix} 1 & 1 & 1 & 3 \\ 0 & -2 & -3 & -8 \\ 0 & 0 & 1 & 2 \end{pmatrix}.$$

Hence

$$\begin{pmatrix} 1 & 1 & 1 & 3 \\ 1 & -1 & -2 & -5 \\ 2 & 3 & 1 & 5 \end{pmatrix} \cong \begin{pmatrix} 1 & 1 & 1 & 3 \\ 0 & -2 & -3 & -8 \\ 0 & 0 & 1 & 2 \end{pmatrix}.$$

The original system is equivalent to the simpler system

$$\begin{aligned} x_1 + x_2 + x_3 &= 3 \\ -2x_2 - 3x_3 &= -8 \\ x_3 &= 2. \end{aligned}$$

The solution can be seen to be $x_3 = 2$, $x_2 = 1$, $x_1 = 0$.

The usual technique in this method is to create zeros in the manner illustrated in this example. The sequence of matrix transformations involves systematically creating zeros below the first element in the first row, then below the second element in the second row, etc. The first row is used to create the zeros in the first column, the second row to create the zeros in the second column, etc. One aims at constructing a matrix of the form

$$\begin{pmatrix} a_{11} & a_{12} & a_{13} & \dots & a_{1n} & b_1 \\ 0 & a_{22} & a_{23} & \dots & a_{2n} & b_2 \\ 0 & 0 & a_{33} & \dots & a_{3n} & b_3 \\ \vdots & & & & & \\ 0 & 0 & 0 & \dots & a_{mn} & b_m \end{pmatrix}.$$

The following example illustrates what happens when the solution is not unique.

*Example 2*

Solve if possible the system of equations

$$\begin{aligned} x_1 - x_2 + 2x_3 &= 1 \\ 3x_1 - x_2 + 2x_3 &= 3 \\ 2x_1 - x_2 + 2x_3 &= 2. \end{aligned}$$

The augmented matrix is

$$\begin{pmatrix} 1 & -1 & 2 & 1 \\ 3 & -1 & 2 & 3 \\ 2 & -1 & 2 & 2 \end{pmatrix},$$

and

$$\begin{pmatrix} 1 & -1 & 2 & 1 \\ 3 & -1 & 2 & 3 \\ 2 & -1 & 2 & 2 \end{pmatrix} \underset{\text{Row } 2-3 \text{ (Row 1)}}{\cong} \begin{pmatrix} 1 & -1 & 2 & 1 \\ 0 & 2 & -4 & 0 \\ 2 & -1 & 2 & 2 \end{pmatrix}$$

$$\underset{\text{Row } 3-2 \text{ (Row 1)}}{\cong} \begin{pmatrix} 1 & -1 & 2 & 1 \\ 0 & 2 & -4 & 0 \\ 0 & 1 & -2 & 0 \end{pmatrix} \underset{\text{Row } 3-1/2 \text{ (Row 2)}}{\cong} \begin{pmatrix} 1 & -1 & 2 & 1 \\ 0 & 2 & -4 & 0 \\ 0 & 0 & 0 & 0 \end{pmatrix}.$$

Hence the original system is equivalent to the system

$$\begin{aligned} x_1 - x_2 + 2x_3 &= 1 \\ x_2 - 2x_3 &= 0. \end{aligned}$$

The solutions are of the form $x_1 = 1$, $x_2 = 2c$, $x_3 = c$, $c$ an arbitrary constant.

We know that systems of equations do not necessarily have solutions. When the method of this section is applied to a system that has no solutions this fact becomes apparent as the following example illustrates.

*Example 3*

Solve if possible the system

$$\begin{aligned} x_1 + 2x_2 + 3x_3 &= 3 \\ 2x_1 + x_2 + 3x_3 &= 3 \\ x_1 + x_2 + 2x_3 &= 0. \end{aligned}$$

The augmented matrix is

$$\begin{pmatrix} 1 & 2 & 3 & 3 \\ 2 & 1 & 3 & 3 \\ 1 & 1 & 2 & 0 \end{pmatrix}.$$

We have that

$$\begin{pmatrix} 1 & 2 & 3 & 3 \\ 2 & 1 & 3 & 3 \\ 1 & 1 & 2 & 0 \end{pmatrix} \underset{\text{Row } 2-2\,(\text{Row } 1)}{\cong} \begin{pmatrix} 1 & 2 & 3 & 3 \\ 0 & -3 & -3 & -3 \\ 1 & 1 & 2 & 0 \end{pmatrix} \underset{\text{Row } 3-\text{ Row } 1}{\cong} \begin{pmatrix} 1 & 2 & 3 & 3 \\ 0 & -3 & -3 & -3 \\ 0 & -1 & -1 & -3 \end{pmatrix}$$

$$\underset{\text{Row } 3-1/3\text{ Row }(2)}{\cong} \begin{pmatrix} 1 & 2 & 3 & 3 \\ 0 & -3 & -3 & -3 \\ 0 & 0 & 0 & -2 \end{pmatrix}.$$

Hence the original system is equivalent to the system

$$\begin{aligned} x_1 + 2x_2 + 3x_3 &= 3 \\ -3x_2 - 3x_3 &= -3 \\ 0x_1 + 0x_2 + 0x_3 &= -2. \end{aligned}$$

This system has no solutions, hence the original system has no solutions.

The following example illustrates a further situation that can arise.

*Example 4*

Solve if possible the system

$$\begin{aligned} x_1 - x_2 + 2x_3 &= 3 \\ 2x_1 - 2x_2 + 5x_3 &= 4 \\ x_1 + 2x_2 - x_3 &= 2 \\ x_2 + 2x_3 &= 1 \end{aligned}$$

$$\begin{pmatrix} 1 & -1 & 2 & 3 \\ 2 & -2 & 5 & 4 \\ 1 & 2 & -1 & 2 \\ 0 & 1 & 2 & 1 \end{pmatrix} \underset{\text{Row } 2-2\,(\text{Row } 1)}{\cong} \begin{pmatrix} 1 & -1 & 2 & 3 \\ 0 & 0 & 1 & -2 \\ 1 & 2 & -1 & 2 \\ 0 & 1 & 2 & 1 \end{pmatrix}$$

$$\underset{\text{Row } 3-\text{ Row } 1}{\cong} \begin{pmatrix} 1 & -1 & 2 & 3 \\ 0 & 0 & 1 & -2 \\ 0 & 3 & -3 & -1 \\ 0 & 1 & 2 & 1 \end{pmatrix}.$$

One cannot at this stage use the second row to "get rid" of the 3 and 1 in the second column because the element in the $a_{22}$ slot is 0. This difficulty can be overcome by interchanging the second and third rows:

$$\underset{\substack{\text{interchange}\\ \text{Rows 2 and 3}}}{\cong}\begin{pmatrix}1 & -1 & 2 & 3\\ 0 & 3 & -3 & -1\\ 0 & 0 & 1 & -2\\ 0 & 1 & 2 & 1\end{pmatrix}\underset{3(\text{Row } 4)-\text{Row } 2}{\cong}\begin{pmatrix}1 & -1 & 2 & 3\\ 0 & 3 & -3 & -1\\ 0 & 0 & 1 & 2\\ 0 & 0 & 9 & 4\end{pmatrix}$$

$$\underset{\text{Row } 4-9\text{ Row } 3}{\cong}\begin{pmatrix}1 & -1 & 2 & 3\\ 0 & 3 & -3 & -1\\ 0 & 0 & 1 & 2\\ 0 & 0 & 0 & -14\end{pmatrix}.$$

Hence the system is equivalent to the system

$$\begin{aligned} x_1 - x_2 + 2x_3 &= 3\\ 3x_2 - 3x_3 &= -1\\ x_3 &= 2\\ 0x_1 + 0x_2 + 0x_3 &= -14, \end{aligned}$$

and thus has no solution.

## Exercises

Solve if possible:

1 $\begin{aligned} 2x_1 + x_2 + x_3 &= 0\\ x_1 - x_2 + x_3 &= 1\\ 3x_1 + 2x_2 + 2x_3 &= 1 \end{aligned}$

2 $\begin{aligned} x_1 + x_2 + x_3 &= 6\\ x_1 - x_2 + x_3 &= 2\\ x_1 + 2x_2 + 3x_3 &= 14 \end{aligned}$

3 $\begin{aligned} 2x_1 - x_2 - x_3 &= 2\\ x_1 - x_2 + x_3 &= 2\\ 3x_1 - 2x_2 + x_3 &= 5 \end{aligned}$

4 $\begin{aligned} x_1 - x_2 \qquad &= 0\\ x_2 + x_3 &= 1\\ 2x_1 + 3x_2 + 5x_3 &= 4 \end{aligned}$

5 $\begin{aligned} 2x_1 + x_2 + x_3 &= 3\\ x_1 - x_2 + 2x_3 &= 0\\ 3x_1 + x_2 + x_3 &= 2 \end{aligned}$

6 $$\begin{aligned} x_1 - x_2 + x_3 &= 1 \\ 2x_1 - 2x_2 + 3x_3 &= 3 \\ x_1 - x_2 - x_3 &= -1 \end{aligned}$$

7 $$\begin{aligned} x_1 + 2x_2 - x_3 + x_4 &= 4 \\ 2x_1 + 3x_2 + x_3 - 2x_4 &= 6 \\ x_1 + x_2 + 3x_3 + 2x_4 &= 5 \\ 3x_1 + x_2 - x_3 - 3x_4 &= 4 \end{aligned}$$

8 $$\begin{aligned} x_1 + x_2 + x_3 + x_4 &= 3 \\ x_1 - x_2 - 2x_3 + 3x_4 &= 4 \\ 2x_1 + x_2 + x_3 + 3x_4 &= 7 \\ x_1 + 2x_2 - x_3 - 2x_4 &= 5 \end{aligned}$$

9 $$\begin{aligned} x_1 - x_2 + 2x_3 \qquad &= 1 \\ 2x_1 - 2x_2 + 2x_3 + 3x_4 &= 2 \\ x_1 + 2x_3 \qquad - 3x_4 &= 1 \\ 3x_1 + x_2 - 4x_3 + 3x_4 &= 3 \end{aligned}$$

10 Prove that when a system of equations is changed by multiplying one equation by a constant and adding it to another equation the solutions remain unchanged.

11 Prove that if a system of linear equations has no solutions then any system obtained from this system through elementary transformations has no solutions.

12 Prove theorem 2 of this section.

## 5.8 Cramer's Rule

Consider the following system of $n$ equation in $n$ unknowns, which is assumed to have a non-singular matrix of coefficients.

$$\begin{aligned} a_{11}x_1 + a_{12}x_2 + \cdots a_{1n}x_n &= y_1 \\ &\vdots \\ a_{n1}x_1 + n_{n2}x_2 + \cdots a_{nn}x_n &= y_n. \end{aligned}$$

Since the matrix of coefficients is non-singular its rank is $n$. Hence a unique solution exists to this system. Here we give a method of determining the solution called *Cramer's Rule*.

*Theorem*

The above system has a unique solution given by

$$x_k = \frac{1}{D(A)} \begin{vmatrix} a_{11} & \dots & a_{1k-1} & y_1 & a_{1k+1} & \dots & a_{1n} \\ \vdots & & \vdots & \vdots & \vdots & & \vdots \\ a_{n1} & \dots & a_{nk-1} & y_n & a_{nk+1} & \dots & a_{nn} \end{vmatrix}, \text{ for } k = 1, 2, \dots, n.$$

Proof: The determinant on the righthand side is the determinant of the matrix obtained by replacing the $k$th column of $A$ by $Y$. This determinant can be manipulated using the multilinear properties of determinants:

$$
\begin{aligned}
D(A^1, \ldots, Y, \ldots, A^n) &= D(A^1, \ldots, x_1 A^1 \ldots + x_n A^n, \ldots, A^n) \\
&= D(A^1, \ldots, x_1 A^1, \ldots, A^n) + \ldots D(A^1, \ldots, x_n A^n, \ldots, A^n) \\
&= x_1 D(A^1, \ldots, A^1, \ldots, A^n) + \ldots x_n D(A^1, \ldots, A^n, \ldots, A^n) \\
&= x_k D(A^1, \ldots, A^k, \ldots, A^n), \text{ (This is the only non-zero determinant since all the others have two columns identical),} \\
&= x_k D(A).
\end{aligned}
$$

This proves that the $x_k$ given above is a solution to the system; hence it must be the unique solution.

Cramer's Rule is useful in theoretical work involving systems of linear equations in that it does give an explicit formula for the solution. However it is not usually used to obtain numerical solutions to large systems of equations as the work involved in evaluating large determinants is enormous. Other more efficient methods involving elimination techniques do exist.[1]

## Exercises

Solve, if possible, each of the following systems of linear equations using Cramer's Rule:

1
$$
\begin{aligned}
x + y + z &= 0 \\
3x - 2y + 5z &= -10 \\
2x + 4y + 8z &= 4.
\end{aligned}
$$

2
$$
\begin{aligned}
2x_1 + 2x_2 - x_3 &= 11 \\
-x_1 + x_2 + 2x_3 &= -3 \\
3x_1 - 3x_2 + 2x_3 &= 1.
\end{aligned}
$$

3
$$
\begin{aligned}
x + y + z &= 11 \\
x - 2y + 6z &= 0 \\
2x + 3y + 4z &= 0.
\end{aligned}
$$

4
$$
\begin{aligned}
x - y + w &= 3 \\
y + z - w &= -3 \\
x - 2y + 3w &= 0 \\
x - 3z + 2w &= 6.
\end{aligned}
$$

The following question illustrates how a problem in business reduces to finding a solution to a system of linear equations.

---

[1] See for example: J. K. FADDEEV and V. N. FADDEEVA, *Computational Methods of Linear Algebra* (W. H. Freeman and Company, 1963).

5 Three products $A$, $B$ and $C$ are prepared. 100 tons of a mixture of the products is required. The products cost \$ 1, \$ 9 and \$ 5 per ton, respectively, to prepare. \$ 500 is available to spend. The times involved in preparing the products are 10, 15 and 10 minutes per ton, respectively. A total amount of 1200 minutes is available for the preparation. Find the amounts of each of the products that will be contained in the mixture, by reducing the problem to that of solving a set of three linear equations.

## 5.9 Inverse of a Matrix

We recall from section 2.2 that a square matrix $A$ is invertible if and only if there exists a matrix $B$ such that $AB = BA = I_n$. Here $I_n$ is the multiplicative identity of the set of $n \times n$ matrices. $B$ is called the inverse of $A$ and is denoted $A^{-1}$. In this section a criterion for the existence of an inverse is found and a method for determining the inverse given.

*Theorem*

A square $n \times n$ matrix is invertible if and only if it is non-singular.
Proof: First of all assume that $A$ is invertible. Hence the inverse $A^{-1}$ of $A$ exists and $AA^{-1} = I_n$.

Taking the determinant, $D(AA^{-1}) = 1$, giving $D(A)\,D(A^{-1}) = 1$, hence proving that $D(A) \neq o$ and that $A$ is non-singular.

To prove the converse we associate with the non-singular matrix $A$ the system of linear equations $AX = Y$ and apply Cramer's Rule. Here $X$ and $Y$ are vectors with $n$ components from the underlying field of scalars $\mathbf{F}$. The system has a unique solution given by

$$x_1 = 1/D(A)\,D(Y, A^2, \ldots, A^n), \ldots, x_n = 1/D(A)\,D(A^1, \ldots, A^{n-1}, Y).$$

Expanding $D(Y, A^2, \ldots, A^n)$ in terms of the first column, $D(A^1, Y, \ldots, A^n)$ in terms of the second column etc., the solution may be expressed

$$x_1 = 1/D(A)\,(y_1 D(A_{11}) + \ldots (-1)^{n+n}\, y_n D(A_{n1})), \ldots,$$
$$x_n = 1/D(A)\,((-1)^{1+n}\, y_1 D(A_{1n}) \ldots (-1)^{n+n}\, y_n D(A_{nn})),$$

where $A_{ij}$ is the minor of the element $a_{ij}$ of $A$.

This solution may be expressed as a single vector equation

$$X = BY$$

where

$$B = \frac{1}{D(A)} \begin{pmatrix} D(A_{11}) \ldots (-1)^{n+1}\, D(A_{n1}) \\ \vdots \qquad\qquad \vdots \\ (-1)^{1+n}\, D(A_{1n}) \ldots (-1)^{n+n} D(A_{nn}) \end{pmatrix} \tag{1}$$

Combining the vector equations $AX = Y$ and $X = BY$ by inserting the expression for $X$ from the latter into the first we get

$$ABY = Y$$

Similarly by substituting for $Y$ from the first vector equation into the second,

$$BAX = X$$

Since $X$ and $Y$ were arbitrary this implies that $AB = BA = I_n$, the identity $n \times n$ matrix, and hence that $B = A^{-1}$.

The proof of this theorem also gives us a method for determining the inverse of a matrix. The inverse of a matrix $A$ is given by (1). The matrix

$$\begin{pmatrix} D(A_{11}) & \dots & (-1)^{n+1} D(A_{n1}) \\ \vdots & & \vdots \\ (-1)^{1+n} D(A_{1n}) & \dots & (-1)^{n+n} D(A_{nn}) \end{pmatrix}$$

is called the *adjugate of A* and denoted *adj A*.

Hence $A^{-1} = 1/D(A)\ adj\ A$.

This method for computing inverses of matrices is a practical one for small matrices.

*Example*

Determine the inverse of the matrix $A = \begin{pmatrix} 1 & 2 \\ 3 & 4 \end{pmatrix}$.

$D(A) = 4 - 6 = -2 \neq 0$. Hence $A^{-1}$ exists.

We have that $D(A_{11}) = 4$, $D(A_{12}) = 3$, $D(A_{21}) = 2$, $D(A_{22}) = 1$.

Hence

$$A^{-1} = -1/2 \begin{pmatrix} 4 & -2 \\ -3 & 1 \end{pmatrix}.$$

As a check it can be seen that

$$-1/2 \begin{pmatrix} 1 & 2 \\ 3 & 4 \end{pmatrix} \begin{pmatrix} 4 & -2 \\ -3 & 1 \end{pmatrix} = -1/2 \begin{pmatrix} 4 & -2 \\ -3 & 1 \end{pmatrix} \begin{pmatrix} 1 & 2 \\ 3 & 4 \end{pmatrix} = 1.$$

## Exercises

In exercises 1 to 5 determine whether the inverses of the matrices exist and in the cases where the inverses exist determine the inverse.

1 $\begin{pmatrix} 1 & 4 \\ 3 & 2 \end{pmatrix}$.

2 $\begin{pmatrix} 1 & 2 & 3 \\ 0 & 1 & 2 \\ 4 & 5 & 3 \end{pmatrix}$.

3 $\begin{pmatrix} 1 & 2 \\ 2 & 4 \end{pmatrix}$.

4 $\begin{pmatrix} 0 & 3 & 3 \\ 1 & 2 & 3 \\ 2 & 4 & 6 \end{pmatrix}$.

5 $\begin{pmatrix} \sin t & \cos t \\ \cos t & -\sin t \end{pmatrix}$.

6 If $C$ is a non-singular square matrix prove that $D(C)\,D(C^{-1}) = 1$.

## 5.10 Eigenvalues and Eigenvectors of a Matrix

Let $A$ be an $n \times n$ matrix with elements from a field $\mathbf{F}$. A *non-zero* element $X$ of $\mathbf{F}^n$ is said to be an *eigenvector* of $A$ if there exists a scalar $\lambda$ such that

$$AX = \lambda X.$$

If there exists such a $\lambda$ it is uniquely determined, for suppose there exist two such $\lambda$s, $\lambda_1$ and $\lambda_2$. Then $\lambda_1 X = AX = \lambda_2 X$ implying that $\lambda_1 = \lambda_2$. $\lambda$ is called the *eigenvalue* corresponding to the eigenvector $X$ and the set of all eigenvalues is called the *spectrum* of $A$.

Intuitively an eigenvector of $A$ is a vector that remains unchanged or reversed in direction when $A$ operates on the vector space. If the corresponding eigenvalue is 1 then it is unchanged in magnitude also and is then said to be invariant under $A$.

The eigenvalue problem (that is the necessity to calculate eigenvalues and eigenvectors) arises in many applications, for example, in network analysis, structural mechanics and factor analysis. In many cases a knowledge of the eigenstructure is necessary to analyze the differential equations which arise from the appropriate physical model. Digital computer solution of the eigenvalue problem comprises a fascinating part of the subject of numerical analysis.[1] Here we shall show how the eigenvectors and corresponding eigenvalues of a matrix are determined.

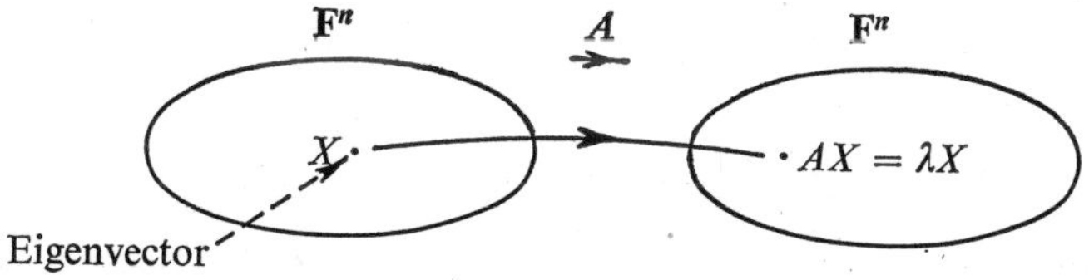


$X$ is an eigenvector with eigenvalue $\lambda$ if and only if $AX = \lambda X$. This vector equation may be written $(A - \lambda I_n) X = O$, which represents the homogeneous set of linear equations

$$\begin{pmatrix} a_{11} - \lambda & \dots & a_{1n} \\ \vdots & & \vdots \\ a_{n1} & \dots & a_{nn} - \lambda \end{pmatrix} \begin{pmatrix} x_1 \\ \vdots \\ x_n \end{pmatrix} = O.$$

Since the augmented matrix and the matrix of coefficients in the above set have the same rank a solution exists. $X = O$ is seen to be a solution. Since we are looking for eigenvectors, which have been defined to be non-

[1] For such considerations the reader is referred to Ben Nobel, *Applied Linear Algebra*, chapters 9–12 (Prentice Hall, 1969).

zero vectors, we are not interested in this solution. Further solutions will exist if and only if the rank of the matrix $(A - \lambda I_n)$ is less than $n$.

The rank of $(A - \lambda I_n)$ is less than $n$ if and only if its determinant is zero. Hence solving the equation $D(A - \lambda I_n) = o$ for $\lambda$ gives the eigenvalues of $A$. The determinantal polynomial $D(A - \lambda I_n)$ is called the *characteristic polynomial* of the matrix and the equation $D(A - \lambda I_n) = o$ is called the *characteristic equation.* The eigenvalues are substituted back into the equation $(A - \lambda I_n) X = 0$ to find the corresponding eigenvectors.

*Example 1.*

Find the eigenvectors and eigenvalues of the matrix $\begin{pmatrix} 1 & 6 \\ 5 & 2 \end{pmatrix}$.

The characteristic polynomial $D(A - \lambda I_2) = o$ is

$$\left| \begin{pmatrix} 1 & 6 \\ 5 & 2 \end{pmatrix} - \lambda \begin{pmatrix} 1 & 0 \\ 0 & 1 \end{pmatrix} \right| = 0.$$

That is

$$\begin{vmatrix} 1 - \lambda & 6 \\ 5 & 2 - \lambda \end{vmatrix} = 0.$$

On expanding, one gets

$$\lambda^2 - 3\lambda - 28 = 0.$$

The roots of this equation are 7 and $-4$ giving the eigenvalues of $A$ to be 7 and $-4$.

The corresponding eigenvectors are now found. For $\lambda = 7$, $(A - \lambda I_2) X = O$ becomes

$$\left[ \begin{pmatrix} 1 & 6 \\ 5 & 2 \end{pmatrix} - 7 \begin{pmatrix} 1 & 0 \\ 0 & 1 \end{pmatrix} \right] \begin{pmatrix} x_1 \\ x_2 \end{pmatrix} = O,$$

which reduces to

$$\begin{pmatrix} -6 & 6 \\ 5 & -5 \end{pmatrix} \begin{pmatrix} x_1 \\ x_2 \end{pmatrix} = O.$$

This gives $x_1 = x_2$. Hence any vector of the type $c(1, 1)$, $c$ being a non-zero scalar, is an eigenvector corresponding to the eigenvalue 7.

For $\lambda = -4$, $(A - \lambda I_2) X = O$ becomes

$$\left[ \begin{pmatrix} 1 & 6 \\ 5 & 2 \end{pmatrix} + 4 \begin{pmatrix} 1 & 0 \\ 0 & 1 \end{pmatrix} \right] \begin{pmatrix} x_1 \\ x_2 \end{pmatrix} = O.$$

This simplifies to

$$\begin{pmatrix} 5 & 6 \\ 5 & 6 \end{pmatrix} \begin{pmatrix} x_1 \\ x_2 \end{pmatrix} = O,$$

implying that $x_1 = (-6/5)\, x_2$. Hence any vector of the type $d(-6/5, 1)$, $d$ being a non-zero scalar, is an eigenvector for $\lambda = -4$.

Note that the eigenvectors in this example can only be determined up to magnitude; for if $AX = \lambda X$, then $A(cX) = \lambda(cX)$ for any non-zero scalar $c$, implying that if $X$ is an eigenvector with eigenvalue $\lambda$, $cX$ is an eigenvector with the same eigenvalue. From this example we might conjecture the result of the following theorem.

*Theorem 1*

Let $A$ be an $n \times n$ matrix over a field $\mathbf{F}$ and $\lambda$ an eigenvalue of $A$. The subset $V_\lambda$ of $\mathbf{F}^n$ consisting of eigenvectors corresponding to $\lambda$ together with the zero vector, is a subspace of $\mathbf{F}^n$.

Proof: Let $X_1$ and $X_2$ be elements of $V_\lambda$. Then

$$AX_1 = \lambda X_1 \quad \text{and} \quad AX_2 = \lambda X_2.$$

Hence

$$AX_1 + AX_2 = \lambda X_1 + \lambda X_2 \Rightarrow A(X_1 + X_2) = \lambda(X_1 + X_2).$$

Thus $X_1 + X_2$ is an eigenvector of $A$ with eigenvalue $\lambda$ and is an element of $V_\lambda$.

Also

$$A(cX_1) = cAX_1 = c\lambda X_1 = \lambda(cX_1),$$

for an arbitrary scalar $c$, giving $cX_1$ to be an element of $V_\lambda$, thus proving the theorem.

The subspace $V_\lambda$ is called the *eigenspace* of $A$ corresponding to the eigenvalue $\lambda$.

*Example 2*

Determine the eigenvalues and eigenspaces of the matrix $\begin{pmatrix} 5 & 4 & 2 \\ 4 & 5 & 2 \\ 2 & 2 & 2 \end{pmatrix}$.

The characteristic polynomial is

$$\left| \begin{pmatrix} 5 & 4 & 2 \\ 4 & 5 & 2 \\ 2 & 2 & 2 \end{pmatrix} - \lambda \begin{pmatrix} 1 & 0 & 0 \\ 0 & 1 & 0 \\ 0 & 0 & 1 \end{pmatrix} \right| = 0.$$

This reduces to $(\lambda - 1)^2 (\lambda - 10) = 0$, giving the eigenvalues to be 10 and 1.

For $\lambda = 10$, the eigenvectors are given by

$$\begin{pmatrix} -5 & 4 & 2 \\ 4 & -5 & 2 \\ 2 & 2 & -8 \end{pmatrix} \begin{pmatrix} x_1 \\ x_2 \\ x_3 \end{pmatrix} = O.$$

This system of linear equations has solutions $x_1 = 2x_3$, $x_2 = 2x_3$, so that the eigenspace is the one-dimensional space $c(2, 2, 1)$. For $\lambda = 1$, the eigenvectors are given by

$$\begin{pmatrix} 4 & 4 & 2 \\ 4 & 4 & 2 \\ 2 & 2 & 1 \end{pmatrix} \begin{pmatrix} x_1 \\ x_2 \\ x_3 \end{pmatrix} = O.$$

This system has solutions $x_3 = -2x_1 - 2x_2$ for arbitary $x_1$ and $x_2$. Hence the eigenspace corresponding to the eigenvalue 1 is the two-dimensional vector space $(x_1, x_2, -2x_1-2x_2)$.

The student may have noticed that 1 is a repeated root of the characteristic polynomial and that the corresponding eigenspace is of dimension 2. It should not be concluded that the dimension of the eigenspace is always the same as the multiplicity of the eigenvalue as a root of the characteristic polynomial. Exercise 5 in the following set illustrates that this need not be so. A further discussion of these concepts is given in chapter 8.

The next theorem deals with the existence of eigenvalues and eigenvectors of a matrix.

*Theorem 2*

If $A$ is a square matrix with elements from the field of complex numbers $C$ there exists an eigenvector of $A$ in $\mathbf{C}^n$.

Proof: Here $A$ is interpreted as being a linear mapping of $\mathbf{C}^n$ onto $\mathbf{C}^n$.

The characteristic polynomial will be over the field of complex numbers. By the fundamental theorem of algebra this polynomial has a root, proving the existence of an eigenvalue. There will exist a corresponding eigenvector, proving the theorem.

It can be seen that one cannot draw up a similar conclusion for square matrices over the field of real numbers. There the fundamental theorem of algebra does not hold and hence there is no guarantee of the existence of eigenvalues and eigenvectors.

If $A$ is a square matrix and $C$ is a non-singular square matrix of the same kind then the mapping $A \mapsto C^{-1}AC$ is called a *similarity transformation.* Two square matrices $A$ and $B$ of the same kind are said to be *similar* if there exists a non-singular matrix $C$ such that $B \mapsto C^{-1}AC$.

*Theorem 3*

Similar matrices have the same characteristic polynomial and hence the same eigenvalues.

Another way of stating this theorem is to say that the eigenvalues of a matrix are invariant under similarity transformations.

Proof: Let $A$ and $B$ similar matrices so that there exists a matrix $C$ such that $B = C^{-1}AC$.

The characteristic polynomial of $B$ is $D(B - \lambda I)$. On substituting for $B$ and using the properties of determinants

$$\begin{aligned} D(B - \lambda I) &= D(C^{-1}AC - \lambda I) = D(C^{-1}(A - \lambda I)\, C) \\ &= D(C^{-1})\, D(A - \lambda I)\, D(C) = D(A - \lambda I). \end{aligned}$$

Hence the characteristic polynomials of $A$ and $B$ are identical; their eigenvalues will then also be identical.

We now discuss the existence of a diagonal matrix similar to a given matrix.

*Theorem 4*

A square $n \times n$ matrix $A$ is similar to a diagonal matrix $B$ if and only if it has $n$ linearly independent eigenvectors. If $A$ is such a matrix the entries of the main diagonal of $B$ are eigenvalues of $A$ and a transforming matrix $C$ has columns consisting of linearly independent eigenvectors of $A$.

Proof: Let $A$ have $n$ linearly independent eigenvectors $v_1, \ldots, v_n$ with corresponding eigenvalues $\lambda_1, \ldots, \lambda_n$. Let $C$ be the matrix having $v_1, \ldots, v_n$ as columns. Write $C = (v_1 \ldots v_n)$. Since $v_1, \ldots, v_n$ are linearly independent then $C$ is non-singular and hence has an inverse. Then

$$\begin{aligned} AC &= A(v_1 \ldots v_n) \\ &= (Av_1 \ldots Av_n) \\ &= (\lambda_1 v_1 \ldots \lambda_n v_n) \\ &= (v_1 \ldots v_n) \begin{pmatrix} \lambda_1 & & O \\ & \ddots & \\ O & & \lambda_n \end{pmatrix} \\ &= C \begin{pmatrix} \lambda_1 & & O \\ & \ddots & \\ O & & \lambda_n \end{pmatrix} \end{aligned}$$

implying that

$$C^{-1}AC = \begin{pmatrix} \lambda_1 & & O \\ & \ddots & \\ O & & \lambda_n \end{pmatrix}.$$

Hence the diagonal matrix $B = \begin{pmatrix} \lambda_1 & & O \\ & \ddots & \\ O & & \lambda_n \end{pmatrix}$ is similar to $A$.

The proof of the converse is obtained by reversing the steps in the above proof.

*Example 3*

Transform, if possible, the matrix $\begin{pmatrix} 5 & 4 & 2 \\ 4 & 5 & 2 \\ 2 & 2 & 2 \end{pmatrix}$ into a diagonal form using a similarity transformation.

This matrix was discussed in example 2. It can be seen that the vector (2, 2, 1) is an eigenvector corresponding to the eigenvalue 10 and that the vectors $(1, 0, -2)$ and $(0, 1, -2)$ are eigenvectors for the eigenvalue 1. The matrix has a set of three linearly independent eigenvectors hence is similar to a diagonal matrix.

Let $C$ be the matrix having these eigenvectors as columns, $C = \begin{pmatrix} 2 & 1 & 0 \\ 2 & 0 & 1 \\ 1 & -2 & -2 \end{pmatrix}$,

Then $C^{-1} = 1/9 \begin{pmatrix} 2 & 2 & 1 \\ 5 & -4 & -2 \\ -4 & 5 & -2 \end{pmatrix}$.

$$C^{-1}AC = 1/9 \begin{pmatrix} 2 & 2 & 1 \\ 5 & -4 & -2 \\ -4 & 5 & -2 \end{pmatrix} \begin{pmatrix} 5 & 4 & 2 \\ 4 & 5 & 2 \\ 2 & 2 & 2 \end{pmatrix} \begin{pmatrix} 2 & 1 & 0 \\ 2 & 0 & 1 \\ 1 & -2 & -2 \end{pmatrix} = \begin{pmatrix} 10 & 0 & 0 \\ 0 & 1 & 0 \\ 0 & 0 & 1 \end{pmatrix}.$$

Hence $A$ can be transformed using a similarity transformation into the diagonal form $\begin{pmatrix} 10 & 0 & 0 \\ 0 & 1 & 0 \\ 0 & 0 & 1 \end{pmatrix}$.

*Theorem 5*

If $A$ is a matrix that can be diagonalized using a similarity transformation then the multiplicity of an eigenvalue as an element of the main diagonal of the diagonal form is its multiplicity as a root of the characteristic equation of $A$.

Proof: Let $B = \begin{pmatrix} \lambda_1 & & 0 \\ & \ddots & \\ 0 & & \lambda_n \end{pmatrix}$ be the diagonal form. Since $A$ and $B$ are similar they have the same characteristic polynomial and hence the same characteristic equation. The characteristic equation of $A$ is thus

$$\begin{vmatrix} \lambda_1 - \lambda & & o \\ & \ddots & \\ o & & \lambda_n - \lambda \end{vmatrix} = o.$$

This reduces to

$$\pi_i (\lambda_i - \lambda)^{k_i} = o$$

where $\pi_i$ indicates product and $k_i$ is the multiplicity of the eigenvalue of $\lambda_i$ as an element of the main diagonal of $B$, proving the theorem.

It can be seen that the matrix of examples 2 and 3 has two eigenvalues, 10 and 1, the first being a single root and the second a double root of the characteristic equation. Hence the diagonal form is $\begin{pmatrix} 10 & 0 & 0 \\ 0 & 1 & 0 \\ 0 & 0 & 1 \end{pmatrix}$.

However, even though the diagonal form can in this manner be written down without having to determine the similarity transformation, a knowledge of the actual transformation is important in many applications of this theory in mathematics, physics and engineering. Such transformations arise for example when co-ordinate transformations are involved in a mathematical model. These are discussed in chapter 6 when one significance of similarity transformations becomes apparent.

Let us now briefly illustrate the use of eigenvalues, eigenvectors and similarity transformations in solving systems of linear differential equations with constant coefficients. If $X'(t) = AX(t)$ is such a system (section 2.2) and if the matrix $A$ has eigenvalues $\lambda_1, \ldots, \lambda_n$ with linearly independent eigenvectors $v_1, \ldots, v_n$ that span the space then it can be shown that $e^{\lambda_1 t}v_1, \ldots, e^{\lambda_n t}v_n$ are $n$ linearly independent solutions. The general solution can be written down $c_1 e^{\lambda_1 t}v_1 + \cdots + c_n e^{\lambda_n t}v_n$ with arbitrary constants $c_1, \ldots, c_n$. If the matrix $A$ does not have eigenvectors that span the space then one has to use generalized eigenvectors to determine the solution.[1] The theory of generalized eigenvectors is presented in chapter 8.

A method that is primarily of theoretical value, involving similarity transformations, is to change the variable using a transformation, $X = CY$, where $C$ is some nonsingular constant matrix. The system then becomes $Y' = C^{-1}ACY$. The object is to choose the matrix $C$ in such a way as to make the matrix $B = C^{-1}AC$ as simple as possible (diagonal if possible). This facilitates the solving of the system $Y' = BY$, and hence of the original system.

## Exercises

Determine the eigenvalues and corresponding eigenspaces of each of the following matrices. Determine a diagonal matrix similar to the given matrix wherever possible, giving the similarity transformation.

1. $\begin{pmatrix} 5 & 4 \\ 1 & 2 \end{pmatrix}$.
2. $\begin{pmatrix} 1 & 0 & 3 \\ 8 & 1 & -1 \\ 5 & 1 & -1 \end{pmatrix}$.

[1] For a discussion of these methods see, for example, F. Brauer and J. A. Nohel, *Qualitative Theory of Ordinary Differential Equations* (W. A. Benjamin Inc., 1969) and D. L. Kreider, R. G. Kuller and D. R. Ostberg, *Elementary Differential Equations* (Addison Wesley, 1968).

3 $\begin{pmatrix} 2 & 1 \\ -1 & 4 \end{pmatrix}$.

4 $\begin{pmatrix} 1 & 6 \\ 5 & 2 \end{pmatrix}$.

5 $\begin{pmatrix} 1 & 2 & 3 \\ 0 & 1 & 2 \\ 0 & 0 & 5 \end{pmatrix}$.

6 Show that if $A$ is a triangular matrix, that is a matrix of the form

$$A = \begin{pmatrix} a_{11} & a_{12} & \cdots & a_{1n} \\ & a_{22} & \cdots & \\ 0 & & \ddots & \\ & & & a_{nn} \end{pmatrix}$$

having all elements on one side of the diagonal zero, the eigenvalues of $A$ are $a_{11}, \ldots, a_{nn}$. Further prove that if the eigenvalues are distinct, corresponding eigenvectors form a basis for $\mathbf{F}^n$.

7 Prove that if $A$ is a square matrix, $A^t$ has the same eigenvalues as $A$.

8 Prove that if $\lambda_1, \ldots, \lambda_n$, are the eigenvalues of $A$, then $c\lambda_1, \ldots, c\lambda_n$ are the eigenvalues of $cA$.

9 If $A$ has eigenvalues $\lambda_1, \ldots, \lambda_n$, prove that $A^2$ has eigenvalues $\lambda_1^2, \ldots, \lambda_n^2$.

10 If $A$ is a non-singular square matrix prove that all its eigenvalues are non-zero. Prove further that the eigenvalues of $A^{-1}$ are the reciprocals of the eigenvalues of $A$.

11 Let $A$, an $n \times n$ matrix with elements from a field $\mathbf{F}$, be interpreted as a linear mapping of $\mathbf{F}^n$ into itself. Assume that $\mathbf{F}^n$ has a basis consisting of eigenvectors of $A$. The characteristic polynomial of $A$ is $p(\lambda) = (\lambda - \lambda_1) \ldots (\lambda - \lambda_n)$, if the eigenvalues of $A$ are $\lambda_1, \ldots, \lambda_n$. We can write a corresponding matrix polynomial $p(A) = (A - \lambda_1 I_n) \ldots (A - \lambda_n I_n)$. Prove that $p(A) = O$; that is, that such a matrix satisfies its own characteristic equation. This result is known as the *Cayley-Hamilton theorem*. It also holds for all $n \times n$ matrices over the field of complex numbers. The student is asked to prove this further result in the exercises of section 8.7. (Hint: Prove that $p(A) X = O$ for arbitrary $X \in \mathbf{F}^n$ by writing $X$ in terms of a basis of eigenvectors.)

12 Prove that distinct eigenspaces intersect in the zero vector.

## CHAPTER 6

# Matrix Representations of Linear Mappings

### 6.1 The Matrix Representation of a Linear Mapping

If $U$ is a vector space of dimension $n$ over a field $\mathbf{F}$ we have seen that there exist isomorphisms between $U$ and $\mathbf{F}^n$ called representation mappings. Any representation mapping enables us to discuss $U$ by discussing $\mathbf{F}^n$. In this chapter we take this concept a stage further. Let $U$ and $V$ be vector spaces over a common field of scalars $\mathbf{F}$, with dimensions $n$ and $m$ respectively. Then $U$ and $V$ can be represented by $\mathbf{F}^n$ and $\mathbf{F}^m$ respectively, using representation mappings. We shall find that we can represent any linear mapping from $U$ to $V$ by a matrix mapping from $\mathbf{F}^n$ to $\mathbf{F}^m$. In this manner any two vector spaces of finite dimensions and their linear mappings can be discussed by means of $n$-tuple, $m$-tuples and matrices.

Let $U$ and $V$ be vector spaces of dimensions $n$ and $m$ respectively over a common field of scalars $\mathbf{F}$, with bases $\{u_1, \ldots, u_n\}$ and $\{v_1, \ldots, v_m\}$. Let $f$ be a linear mapping of $U$ into $V$. Using these bases and the corresponding representation mappings $I_1$ and $I_2$ the spaces $U$ and $V$ have representations $\mathbf{F}^n$ and $\mathbf{F}^m$ respectively. $I_1$ and $I_2$ are defined as follows.

For $u \varepsilon U$, let $u = \sum_{i=1}^{n} b_i u_i$ and for $v \varepsilon V$ let $v = \sum_{j=1}^{m} c_j v_j$. Then

$$I_1(u) = (b_1, \ldots, b_n) \quad \text{and} \quad I_2(v) = (c_1, \ldots, c_m).$$

The following diagram illustrates the mathematical structures defined thus far.

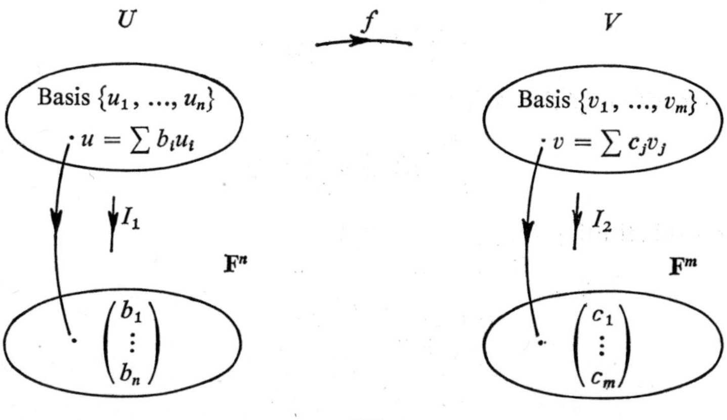

In this chapter the elements of $\mathbf{F}^n$ and $\mathbf{F}^m$ will be interpreted as column vectors; one can pre-multiply an element of $\mathbf{F}^n$ by an $m \times n$ matrix to get an element of $\mathbf{F}^m$. However, for reason of space, we often write these vectors as row vectors in the text.

Since $\mathbf{F}^n$ and $\mathbf{F}^m$ are vector spaces, there exist linear mappings between them. As was illustrated in example 4, section 3.2, matrices can be interpreted as linear mappings from $\mathbf{F}^n$ to $\mathbf{F}^m$. The aim of this section is to construct a matrix $A$, which when used as a linear mapping between $\mathbf{F}^n$ and $\mathbf{F}^m$ is analogous to the linear mapping $f$ between $U$ and $V$. This matrix will be called the matrix representation of $f$ relative to the bases $\{u_1, \ldots, u_n\}$ and $\{v_1, \ldots, v_m\}$. If $f(u) = v$ in the above diagram, then it will be the matrix $A$ such that $A\begin{pmatrix} b_1 \\ \vdots \\ b_n \end{pmatrix} = \begin{pmatrix} c_1 \\ \vdots \\ c_m \end{pmatrix}$.

As noted previously $f$ is determined if its values on the base vectors of $U$ are known. Let

$$\begin{aligned} f(u_1) &= a_{11}v_1 + \cdots + a_{m1}v_m \\ &\vdots \\ f(u_n) &= a_{1n}v_1 + \cdots + a_{mn}v_m. \end{aligned}$$

That is let

$$f(u_i) = \sum_{j=1}^{m} a_{ji}v_j.$$

Here the matrix of coefficients can be found if $f$ is given. Conversely this set of scalars when used as in the above set of equations determines $f$ uniquely. The reason for labeling the coefficients thus will become apparent. The matrix $A$ which is the transpose of this matrix of coefficients is the *matrix representation* of $f$:

$c_1, \ldots, c_m$ are defined by

$$f(u) = \sum_{j=1}^{m} c_j v_j,$$

that is

$$f\Big(\sum_{i=1}^{n} b_i u_i\Big) = \sum_{j=1}^{m} c_j v_j$$

By the linearity of $f$ this reduces to

$$\sum_{i=1}^{n} b_i f(u_i) = \sum_{j=1}^{m} c_j v_j$$

which on substituting for $f(u_i)$ becomes

$$\sum_{i=1}^{n} b_i \sum_{j=1}^{m} a_{ji} v_j = \sum_{j=1}^{m} c_j v_j$$

$$\sum_{j=1}^{m} \sum_{i=1}^{n} a_{ji} b_i v_j = \sum_{j=1}^{m} c_j v_j.$$

Since the vectors $v_1, \ldots, v_m$ are linearly independent,

$$\sum_{i=1}^{n} a_{ji} b_i = c_j.$$

Hence $A \begin{pmatrix} b_1 \\ \vdots \\ b_n \end{pmatrix} = \begin{pmatrix} c_1 \\ \vdots \\ c_m \end{pmatrix}.$

The significance of this result can be seen from the following *commutative* diagram.

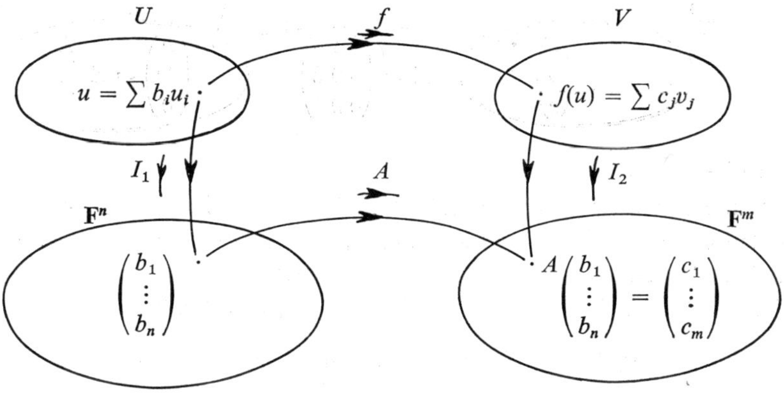


The diagram can also be written in the more compact form

$$\begin{array}{ccc} U & \xrightarrow{\;f\;} & V \\ I_1 \big\downarrow & & \big\downarrow I_2 \\ F^n & \xrightarrow[\;A\;]{} & F^m \end{array}$$

*Example 1*

It was seen in exercise 6, section 1.5 that the set $\{1, x, x^2\}$, the elements being interpreted as functions on the interval [0, 1], can be considered as a set of base vectors for a real three dimensional vector space $U$. Let $D = d/dx$ be an operator on $U$. Find the matrix representation of $D$ relative to this basis.

We determine the effect of $D$ on the base vectors,

$$\begin{aligned} D(1) &= 0 &&= 0(1) + 0(x) + 0(x^2), \\ D(x) &= 1 &&= 1(1) + 0(x) + 0(x^2), \\ D(x^2) &= 2x &&= 0(1) + 2(x) + 0(x^2). \end{aligned}$$

The matrix of coefficients is $\begin{pmatrix} 0 & 0 & 0 \\ 1 & 0 & 0 \\ 0 & 2 & 0 \end{pmatrix}$, hence the matrix representation of $D$ is the transpose of this matrix, namely $\begin{pmatrix} 0 & 1 & 0 \\ 0 & 0 & 2 \\ 0 & 0 & 0 \end{pmatrix}$. The commutative diagram is

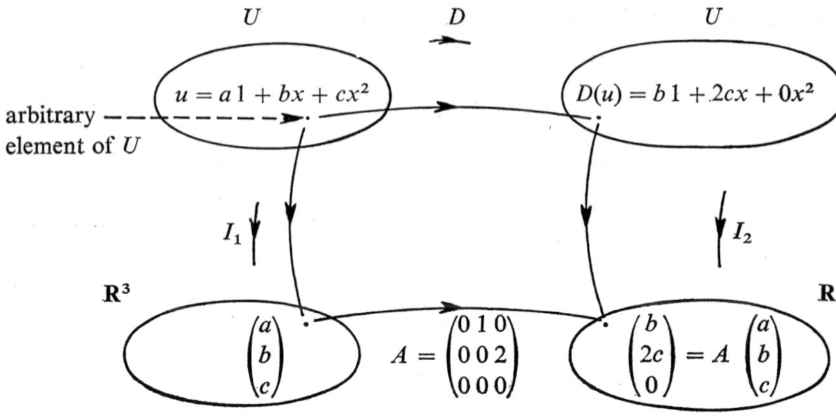


*Example 2*

Determine the matrix representation of the projection operator $f$ of $\mathbf{R}^3$ into $\mathbf{R}^3$ defined by $f: (x, y, z) \mapsto (x, y, 0)$, relative to the usual basis of $\mathbf{R}^3$. (This is the operator of example 2, section 3.2.)

The usual basis for $\mathbf{R}^3$ is the set $\{(1, 0, 0), (0, 1, 0), (0, 0, 1)\}$ and we have

$$f(1, 0, 0) = (1, 0, 0) = 1(1, 0, 0) + 0(0, 1, 0) + 0(0, 0, 1)$$

$$f(0, 1, 0) = (0, 1, 0) = 0(0, 0, 1) + 1(0, 1, 0) + 0(0, 0, 1)$$

$$f(0, 0, 1) = (0, 0, 0) = 0(0, 0, 1) + 0(0, 1, 0) + 0(0, 0, 1)$$

Hence the matrix of coefficients is $\begin{pmatrix} 1 & 0 & 0 \\ 0 & 1 & 0 \\ 0 & 0 & 0 \end{pmatrix}$ giving that the matrix representation of $f$ relative to this basis is $\begin{pmatrix} 1 & 0 & 0 \\ 0 & 1 & 0 \\ 0 & 0 & 0 \end{pmatrix}$. It can be seen that $\begin{pmatrix} 1 & 0 & 0 \\ 0 & 1 & 0 \\ 0 & 0 & 0 \end{pmatrix} \begin{pmatrix} x \\ y \\ z \end{pmatrix} = \begin{pmatrix} x \\ y \\ 0 \end{pmatrix}$.

Notice that the matrix representation of the operator relative to this basis is a diagonal matrix. In chapter 8 we analyze the properties of such operators. We find the significance of the usual basis for such an operator and understand why a diagonal matrix representation is to be expected.

*Theorem*

Let $f$ be an isomorphism of an $n$-dimensional vector space $U$ onto a vector space $V$, both over the field of scalars $\mathbf{F}$. Then $A$, a matrix representation of $f$ relative to arbitrary bases of $U$ and $V$ can be interpreted as an isomorphism of $\mathbf{F}^n$ onto itself. Further, $A^{-1}$ exists and is the matrix representation of $f^{-1}$ relative to the same bases. ↳ multiplied × A = I

↳ compose with function f: I undoes the F (antiderivative)

Proof: Let $A$ be the matrix representation of $f$ relative to bases $\{u_1, \ldots, u_n\}$ and $\{v_1, \ldots, v_n\}$ of $U$ and $V$ respectively and let the representation isomorphisms be $I_1$ and $I_2$.

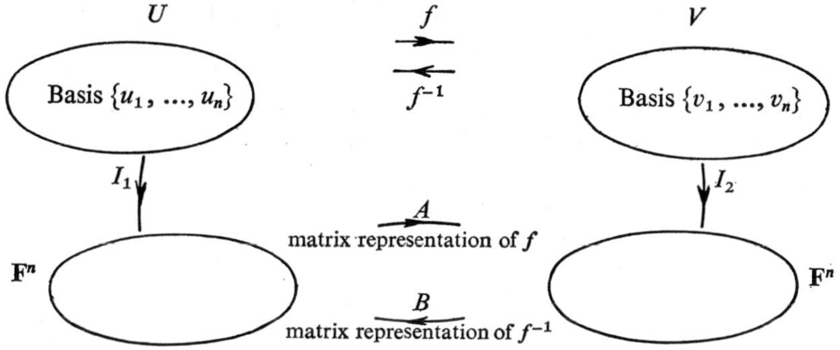


$A$ being a matrix mapping is of course linear. Since $I_1^{-1}$, $f$ and $I_2$ are all isomorphisms, $I_2 \circ f \circ I_1^{-1}$ is an isomorphism of $\mathbf{F}^n$ onto $\mathbf{F}^n$ and from the above commutative diagram it can be seen that $I_2 \circ f \circ I_1^{-1} = A$. Hence $A$, a matrix representation of $f$, is an isomorphism of $\mathbf{F}^n$ onto itself.

Since $f$ is an isomorphism it has an inverse $f^{-1}$. Let $B$ be the matrix representation of $f^{-1}$ relative to the above bases of $U$ and $V$. Then we see from the commutative diagram that $B$ is a mapping of $\mathbf{F}^n$ into itself such that

$$B = I_1 \circ f^{-1} \circ I_2^{-1}.$$

Hence, for elements $X$ and $Y$ of $\mathbf{F}^n$, since $A = I_2 \circ f \circ I_1^{-1}$,

$$ABX = I_2 \circ f \circ I_1^{-1} \circ I_1 \circ f^{-1} \circ I_2^{-1} X = X = I_n X,$$

and

$$BAY = I_1 \circ f^{-1} \circ I_2^{-1} \circ I_2 \circ f \circ I_1^{-1} Y = Y = I_n Y,$$

where $I_n$ is the unit $n \times n$ matrix.

Hence $AB = BA = I_n$ implying that $B = A^{-1}$.

## Exercises

1 Find the matrix representation of the following mapping of $\mathbf{R}^3$ into $\mathbf{R}^2$ relative to the usual bases of these spaces.

$$f(x, y, z) = (x, z)$$

2 Determine the matrix representation of the following operator on $\mathbf{R}^3$, relative to the usual basis for $\mathbf{R}^3$.

$$f\colon (x, y, z) \mapsto (x, 0, 0).$$

This operator is another example of a projection operator. It projects $\mathbf{R}^3$ into a one-dimensional subspace.

3 As in example 1, the set $\{1, 2x, 3x^2\}$ can be considered a set of base vectors for a three-dimensional real vector space $U$. Let $D = d/dx$ be a linear mapping of $U$ into $U$. Find the matrix representation of $D$ relative to this basis.

4 Let $V$ be the real vector space generated by the functions $\sin t$ and $\cos t$ defined on the interval $0 \leqq t \leqq \pi$. If $D$ is the operation of taking the derivative find the matrix representation of the linear differential operator $D^2 + 2D + 1$ of $V$ into $V$ relative to the basis $\{\sin t, \cos t\}$ of $V$.

5 Let $f$ be the linear mapping from a vector space $V$ with basis $\{v_1, v_2, v_3\}$ onto a vector space $W$ with basis $\{w_1, w_2\}$, defined by $f(v_1) = w_1$, $f(v_2) = w_2$, $f(v_3) = O$. Find the matrix representation of $f$ relative to these bases.

## 6.2 Isomorphisms between spaces of Linear Mappings

Let $U$ and $V$ be vector spaces with bases $\{u_1, \ldots, u_n\}$ and $\{v_1, \ldots, v_m\}$ over a common field of scalars $\mathbf{F}$. We have seen that $U$ and $V$ have representations $\mathbf{F}^n$ and $\mathbf{F}^m$ under isomorphisms $I_1$ and $I_2$ relative to these bases and that any linear mapping $f$ has a matrix representation $A$ relative to these bases. Let $L(U; V)$ (section 3.3) represent the vector space of linear mappings from $U$ to $V$. Let $M_{mn}$ represent the set of $m \times n$ matrices with elements from $\mathbf{F}$. Then each matrix representation of an element of $L(U; V)$ is an element of $M_{mn}$. Hence the concept of associating a matrix representation with every linear mapping of $U$ into $V$ is to define a mapping $I_3$ of $L(U; V)$ into $M_{mn}$. We shall now show that this mapping is in fact an isomorphism. Since the elements of $M_{mn}$ can be interpreted as linear mappings of $\mathbf{F}^n$ into $\mathbf{F}^m$ the structures can be illustrated thus:

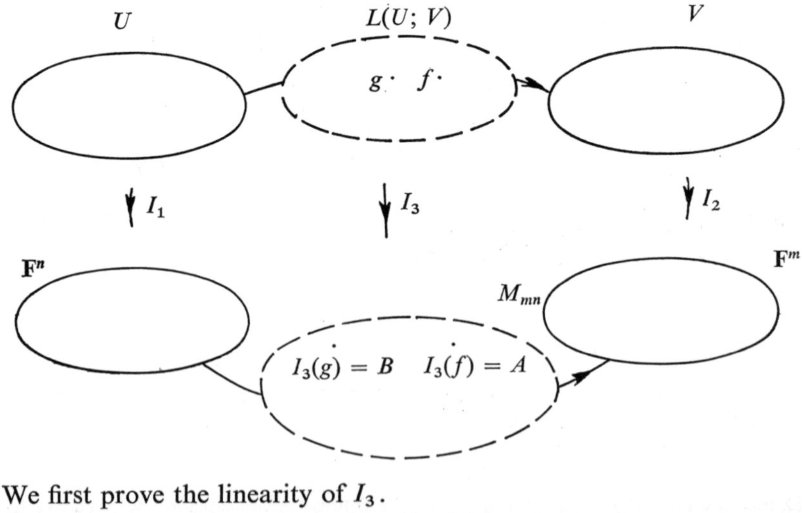


We first prove the linearity of $I_3$.

Let $f$ and $g$ be elements of $L(U; V)$ with representation matrices $A$ and $B$ respectively in $M_{mn}$. Hence $I_3(f) = A$ and $I_3(g) = B$. For the linearity part of the proof we have to show that $I_3(f + g) = I_3(f) + I_3(g)$ and that $I_3(af) = aI_3(f)$, for an arbitrary scalar $a$.

By definition of the matrix representations of linear mappings, if $A = (a_{ij})$ and $B = (b_{ij})$,

$$f(u_i) = \sum_{j=1}^{m} a_{ji}v_j \quad \text{and} \quad g(u_i) = \sum_{j=1}^{m} b_{ji}v_j.$$

These identities give

$$(f + g)(u_i) = f(u_i) + g(u_i) = \sum_{j=1}^{m} (a_{ji} + b_{ji})\, v_j$$

and the matrix representation of $f + g$ is $A + B$.

Therefore $I_3(f + g) = A + B = I_3 f + I_3 g$.

Since $f(u_j) = \sum_{j=1}^{m} a_{ji}v_j$, $af(u_i) = \sum_{j=1}^{m} aa_{ji}v_j$ for an arbitrary scalar $a$.

This implies that the matrix representation of the linear mapping $af$ is $(aa_{ij})$. By the rule for scalar multiplication of a matrix this means that the matrix representation of $af$ is $a(a_{ij})$ or $aA$.

Hence $I_3(af) = aA = aI_3 f$.

The mapping $I_3$ is therefore linear.

The next property of $I_3$ to be proved is its injective property. Let $f$ and $g$, with matrix representations $A$ and $B$, be distinct elements of $L(U; V)$. Then since $f$ and $g$ are determined by their effect on the base vectors of $U$ their effect on some base vector say $u_k$ must differ, $f(u_k) \neq g(u_k)$, implying that $\sum_{j=1}^{m} a_{jk}v_j \neq \sum_{j=1}^{m} b_{jk}v_j$. Since the vectors $v_1, \ldots, v_n$ are linearly independent this implies that $a_{pk} \neq b_{pk}$ for some $1 \leqq p \leqq m$. Hence $A \neq B$, implying that if $f \neq g$, $I_3(f) \neq I_3(g)$. Therefore $I_3$ is injective.

Finally we prove that $I_3$ is surjective. Let $A$ be an arbitrary element of $M_{mn}$ and let $h$ be the composite mapping $I_2^{-1} \circ A \circ I_1$ of $U$ into $V$. Each of $I_1$, $A$ and $I_2^{-1}$ is linear hence $h$ is linear and an element of $L(U; V)$. Furthermore $A$ is the matrix representation of $h$. Therefore $I_3(h) = A$, proving that $I_3$ is surjective.

Hence the mapping defined by matrix representation is an isomorphism of $L(U; V)$ onto $M_{mn}$.

In particular this mapping is an isomorphism of $L(\mathbf{F}^n; \mathbf{F}^m)$ onto $M_{mn}$. Hence we can think of the vector space $M_{mn}$ as being the set of linear mappings of $\mathbf{F}^n$ onto $\mathbf{F}^m$—their mathematical properties are identical. In future, in this context we shall use the notation $L(\mathbf{F}^n; \mathbf{F}^m)$ for the set of matrix representations.

## Exercises

1 In section 2.2 we defined an operation of multiplication between matrices of certain types and discussed the fact that the set of $n \times n$ matrices with this additional operation had more structure than a vector space—it formed an algebra. In a similar manner a further operation of composition can be defined on the vector space $L(U; U)$

of operators on $U$, making it into an algebra. Consider the isomorphism $I_3$ of $L(U; U$ onto $M_{nn}$ discussed in this section and prove that it preserves the operations of composition and matrix multiplication in the following sense:

$$\text{if } I_3(f) = A \quad \text{and} \quad I_3(g) = B,$$

then

$$I_3(g \circ f) = BA .$$

Hence $I_3$ preserves the algebra structures of $L(U; U)$ and $M_{nn}$.

2 Let $U$, $V$ and $W$ be three finite-dimensional vector spaces and $f$ a linear mapping of $U$ into $V$, $g$ a linear mapping of $V$ into $W$. If $f$ and $g$ have matrix representations $A$ and $B$ relative to given bases prove that $g \circ f$ has matrix representation $BA$ relative to those same bases of $U$ and $W$.

## 6.3 Representations of Vectors Relative to Different Bases

We have seen that if $U$ is a vector space of dimension $n$ over a field $\mathbf{F}$ there exsist many isomorphisms of $U$ onto $\mathbf{F}^n$; in fact an isomorphism corresponds to each basis of $U$. We have called such isomorphisms representation mappings. Each $\mathbf{F}^n$, considered as the image of a representation mapping, is said to be a representation of $U$. The purpose of this section is to demonstrate how the representations of $U$ are related.

Let $\{u_1, ..., u_n\}$ and $\{u'_1, ..., u'_n\}$ be two bases for $U$. Let $I$ be the identity mapping of $U$ onto $U$ and let $I_1$ and $I_2$ be the representation mappings of $U$ onto $\mathbf{F}^n$ corresponding to the two bases. If $u$ is an arbitrary element of $U$ then there exist two sets of scalars $\{a_1, ..., a_n\}$ and $\{a'_1, ..., a'_n\}$ such that

$$u = \sum_{i=1}^{n} a_i u_i \quad \text{and} \quad u = \sum_{i=1}^{n} a'_i u'_i .$$

Hence $I_1(u) = (a_1, ..., a_n)$, $I_2(u) = (a'_1, ..., a'_n)$ and $(a_1, ..., a_n)$, $(a'_1, ..., a'_n)$ are the representations of $u$ relative to the two bases. We shall show how these representations are related.

Let $A$ be the matrix representation of the indentity mapping from $U$ onto $U$ relative to these two bases.

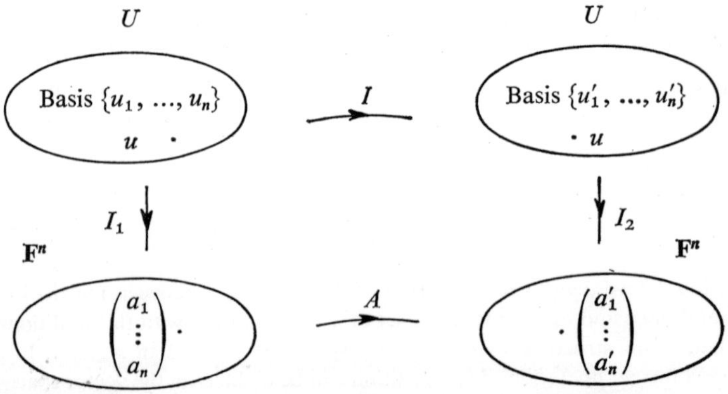

We know that the diagram is a commutative diagram. Hence

$$A\begin{pmatrix}a_1\\ \vdots\\ a_n\end{pmatrix} = I_2 \circ I \circ I_1^{-1}\begin{pmatrix}a_1\\ \vdots\\ a_n\end{pmatrix} = \begin{pmatrix}a_1'\\ \vdots\\ a_n'\end{pmatrix},$$

giving

$$\begin{pmatrix}a_1'\\ \vdots\\ a_n'\end{pmatrix} = A\begin{pmatrix}a_1\\ \vdots\\ a_n\end{pmatrix}$$

The representations of $u$ relative to the distinct bases are thus related in this manner through the matrix representation of the identity mapping relative to these bases.

*Example*

Let $U$ be a real vector space with basis $\{u_1, u_2, u_3\}$. Then $\{u_1 + u_2, u_1 - u_2, u_3\}$ is a second basis. If the representation of a vector $u$ relative to the first basis is (1, 2, 3) find its representation relative to the second basis.

We have that $u = u_1 + 2u_2 + 3u_3$, and we want to find scalars $a$, $b$ and $c$ such that $u = a(u_1 + u_2) + b(u_1 - u_2) + cu_3$.

First we need the representation of the identity mapping relative to these bases. We have

$$I(u_1) = u_1 = \tfrac{1}{2}(u_1 + u_2) + \tfrac{1}{2}(u_1 - u_2) + 0u_3,$$

$$I(u_2) = u_2 = \tfrac{1}{2}(u_1 + u_2) - \tfrac{1}{2}(u_1 - u_2) + 0u_3,$$

$$I(u_3) = u_3 = 0(u_1 + u_2) + 0(u_1 - u_2) + 1u_3.$$

Hence the matrix representation of $I$ is $\begin{pmatrix}\frac{1}{2} & \frac{1}{2} & 0\\ \frac{1}{2} & -\frac{1}{2} & 0\\ 0 & 0 & 1\end{pmatrix}$.

The representation of $u$ relative to the second basis is therefore given by

$$\begin{pmatrix}\frac{1}{2} & \frac{1}{2} & 0\\ \frac{1}{2} & -\frac{1}{2} & 0\\ 0 & 0 & 1\end{pmatrix}\begin{pmatrix}1\\ 2\\ 3\end{pmatrix} = \begin{pmatrix}\frac{3}{2}\\ -\frac{1}{2}\\ 3\end{pmatrix}$$

and

$$u = \tfrac{3}{2}(u_1 + u_2) - \tfrac{1}{2}(u_1 - u_2) + 3u_3.$$

The commutative diagram is

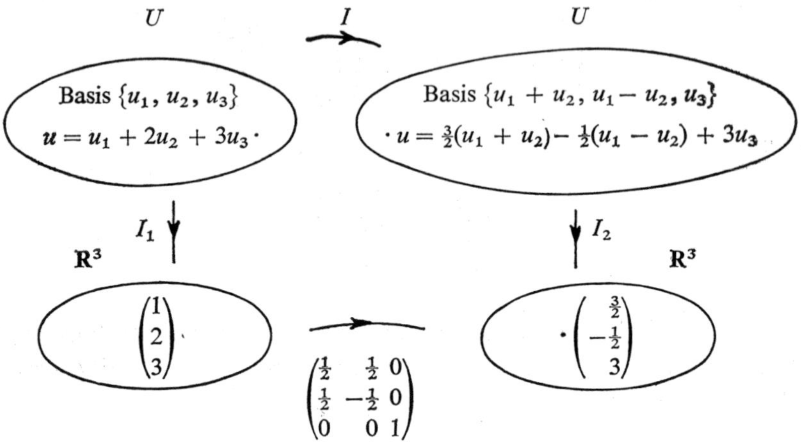


## Exercises

1 Let $U$ be a real vector space with basis $\{u_1, u_2, u_3\}$. Then $\{u_1 + 2u_2, u_2 - u_1, u_3\}$ is a second basis. If the representation of a vector $u$ relative to the first basis is (2, 3, 4) find its representation relative to the second basis.

2 The sets $\{(1, 0, 0), (0, 1, 0), (0, 0, 1)\}$ and $\{(1, 2, 3), (0, 1, 5), (0, 0, 4)\}$ are bases for $\mathbf{R}^3$. If $u$ is the vector (0, 1, 0) relative to the first basis what is its representation relative to the second basis?

3 The sets $\{1, x, x^2\}$ and $\{2, x^2 - x, x^2 - 1\}$ are bases for the same vector space, a subspace of the space of differentiable functions on the interval [0, 1]. Let $y$ be an element of this space with representation $(a, b, c)$ relative to the first basis. What is the representation of $y$ relative to the second basis?

4 Prove that the matrix representation of the identity operator mapping $\mathbf{F}^n$ with the usual basis into $\mathbf{F}^n$ with the basis $\{v_1, \ldots, v_n\}$ is the matrix having as columns the vectors $v_1, \ldots, v_n$.

## 6.4 Change of Matrix Representation with Change of Bases

If $U$ and $V$ are vector spaces over a common field $\mathbf{F}$ and $f$ is a linear mapping of $U$ into $V$ we know that we can find matrix representations of $f$ relative to different bases of $U$ and $V$. Corresponding to each basis for $U$ and $V$ there will exist a matrix representation of $f$. The purpose of this section is to show how the matrix representations of $f$ corresponding to different bases are related.

Let $A$ and $B$ be one pair of bases for $U$ and $V$ respectively and $A'$ and $B'$ a second pair. Let $M_B^A(f)$ and $M_{B'}^{A'}(f)$ be the matrix representations of $f$ relative to these pairs.

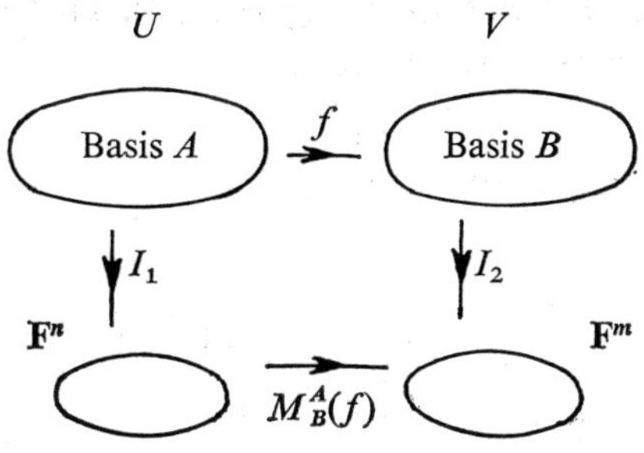


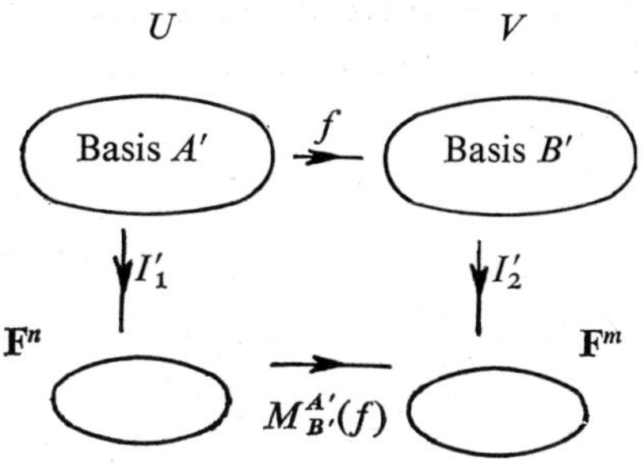


Here $I_1$, $I_2$ and $I_1'$, $I_2'$ are the usual representation mappings, corresponding to the bases $A$, $B$, $A'$ and $B'$ respectively.

We wish to determine the manner in which $M_B^A(f)$ and $M_{B'}^{A'}(f)$ are related. We shall be guided by commutative diagrams. The two diagrams above can be put together to form the one diagram:

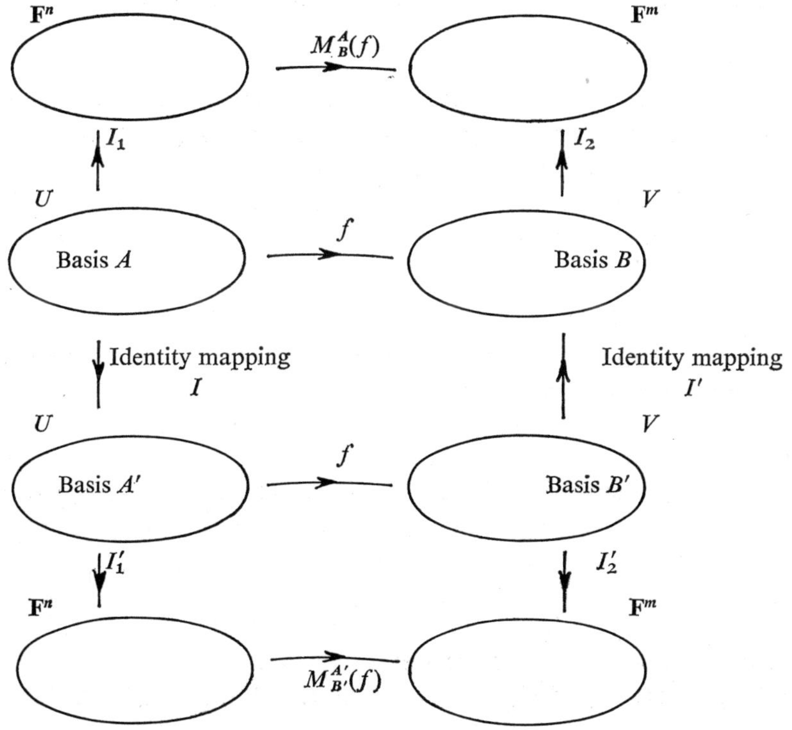


Here the top half consists of the first diagram (inverted) and the bottom half consists of the second diagram.

The mapping of $\mathbf{F}^n$ into $\mathbf{F}^n$ defined by $I_1' \circ I \circ (I_1)^{-1}$ is the matrix representation of $I$ relative to the bases $A$ and $A'$. Call this mapping $M_{A'}^A(I)$. Similarly the mapping of $\mathbf{F}^m$ into $\mathbf{F}^m$ defined by $I_2 \circ I' \circ (I_2')^{-1}$ is the matrix representation of $I'$ relative to the bases $B'$ and $B$. Call this mapping $M_B^{B'}(I')$. This diagram then reduces to the following commutative diagram.

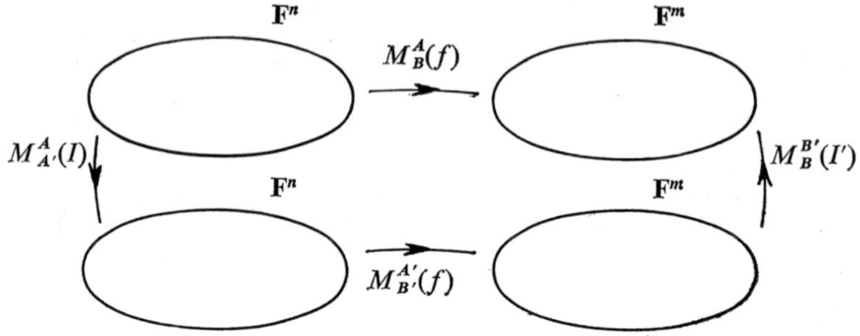


Hence

$$M_B^A(f) = M_B^{B'}(I') \circ M_{B'}^{A'}(f) \circ M_{A'}^A(I),$$

giving the relation between the two matrix representations of $f$.

*Example 1*

Let $U$ be a vector space with bases $A = \{u_1, \ldots, u_n\}$ and $A' = \{u_1', \ldots, u_n'\}$, and let $I$ be the identity mapping of $U$ onto $U$. Then $M_A^{A'}(I) \circ M_{A'}^A(I) = I_n$, the identity $n \times n$ matrix:

In the previous theory let $U = V$ and $A$, $B$ both be the basis $A$ of $U$, $A'$ and $B'$ both be the basis $A'$ of $U$ and $f$ be the identity mapping $I$ of $U$ onto $U$. Then

$$M_A^A(I) = M_A^{A'}(I) \circ M_{A'}^{A'}(I) \circ M_{A'}^A(\mathrm{I}).$$

But $M_A^A(I) = M_{A'}^{A'}(I) = I_n$, giving the result. Hence the one matrix is the multiplicative inverse of the other.

*Example 2*

Let $U$ be a vector space with bases $A$ and $A'$. Let $f$ be an operator on $U$. Then $M_A^A(f)$ and $M_{A'}^{A'}(f)$, the two matrix representations of $f$ relative to these bases, are related:

$$M_A^A(f) = M_A^{A'}(I) \circ M_{A'}^{A'}(f) \circ M_{A'}^A(I).$$

To see this substitute in the result of the theory.

Further, $M_{A'}^A(I) = (M_A^{A'}(I))^{-1}$, by example 1. Call this matrix $C$. Hence

$$M_A^A(f) = C^{-1} M_{A'}^{A'}(f)\, C.$$

This type of transformation, namely $M_{A'}^{A'}(f) \leftrightarrow M_A^A(f)$, is an example of the similarity transformation that was introduced in section 5.10. Hence two different matrix representations of the operator $f$ are related through a similarity transformation.

At this point one might ask the question: for a linear mapping $f$ from $U$ into $V$ do there exist bases for $U$ and $V$ such that $f$ has a particularly simple matrix representation? We shall find that certain linear mappings such as projection operators do have representations as diagonal matrices while others do not. The study of this problem is called the "spectral theory of operators." It is the topic of chapter 8. There a method is given for finding the bases that lead to the simplest matrix representation.

## Exercises

1 Draw a commutative diagram to illustrate the result of example 1.

2 Draw a commutative diagram to illustrate the result of example 2.

3 Let $A = \{(1, 0, 0), (0, 1, 0), (0, 0, 1)\}$ and $A' = \{(1, 1, 1), (0, 1, 2), (0, 0, 3)\}$ be a pair of bases for $\mathbf{R}^3$ and $B = \{(1, 0), (0, 1)\}$, $B' = \{(1, 1), (0, 1)\}$ be a pair of bases for $\mathbf{R}^2$. $f$ defined by $f(x, y, z) = (x, y)$ is a linear mapping of $\mathbf{R}^3$ into $\mathbf{R}^2$. Find $M_B^A(f)$ and $M_{A'}^A(I)$, $M_{B'}^{A'}(f)$ and $M_B^{B'}(I')$ and illustrate the result

$$M_B^A(f) = M_B^{B'}(I') \circ M_{B'}^{A'}(f) \circ M_{A'}^A(I)$$

4 If $V$ and $W$ are vector spaces with bases $A$ and $A'$ respectively, and if $f$ and $g$ are linear maps of $V$ into $W$, prove that

$$M_{A'}^A(f+g) = M_{A'}^A(f) + M_{A'}^A(g)$$

5 The set $A = \{1, x, x^2\}$ of functions on the interval $[0, 1]$ generates a three-dimensional real vector space $V$ and $D = d/dx$ is a linear mapping of $V$ into itself. $B = \{2, x - x^2, x + x^2\}$ is a second basis for $V$. Find the matrix representations of $D$ relative to both of these bases and show that the representations are related through a similarity transformation.

6 In example 2 of this section we have proved that two different matrix representations of the operator $f$ are related through a similarity transformation. Prove the converse, namely that two matrices related by a similarity transformation may be interpreted as two representations of a single operator.

## 6.5 Matrix Representations of Bilinear Forms and Quadratic Forms

Having discussed the representations of linear mappings by matrices, we now turn our attention to the possibility of representing bilinear forms by matrices.

Let $V$ and $W$ be vector spaces of dimensions $n$ and $m$ respectively over a common field of scalars $\mathbf{F}$ and let $f$ be a bilinear form on $V \times W$. If the sets $\{v_1, \ldots, v_n\}$ and $\{w_1, \ldots, w_m\}$ are bases for $V$ and $W$, then there exist representation isomorphisms $I_1$ and $I_2$ of $V$ onto $\mathbf{F}^n$ and $W$ onto $\mathbf{F}^m$ respectively, defined by

$$I_1(v) = (a_1, \ldots, a_n), \quad I_2(w) = (b_1, \ldots, b_m),$$

where $v = \sum_{i=1}^{n} a_i v_i$ and $w = \sum_{j=1}^{m} b_j w_j$ are arbitrary vectors of $V$ and $W$. Let

$(a_1, \ldots, a_n) = X$ and $(b_1, \ldots, b_m) = Y$ for convenience. Let $I_3$ be the mapping of $V \times W \to \mathbf{F}^n \times \mathbf{F}^m$ defined by

$$I_3(v, w) = (X, Y).$$

Then $I_3$ is a bijective mapping of $V \times W$ onto $\mathbf{F}^n \times \mathbf{F}^m$. Further $I_3$ can be seen to preserve the vector-space structures of both $V$ and $W$. It is a structure preserving mapping of $V \times W$ onto $\mathbf{F}^n \times \mathbf{F}^m$. Hence $V \times W$ and $\mathbf{F}^n \times \mathbf{F}^m$ are identical from the mathematical point of view under the correspondence defined by $I_3$. The bilinear form $f$ is a mapping of $V \times W$ into $\mathbf{F}$. We shall define in terms of a matrix $A$ a mapping of $\mathbf{F}^n \times \mathbf{F}^m$ into $\mathbf{F}$ that has the same properties relative to $\mathbf{F}^n \times \mathbf{F}^m$ as $f$ has relative to $V \times W$. $A$ will then be a matrix representation of $f$.

Hence the problem confronting us in terms of a commutative diagram is the following:

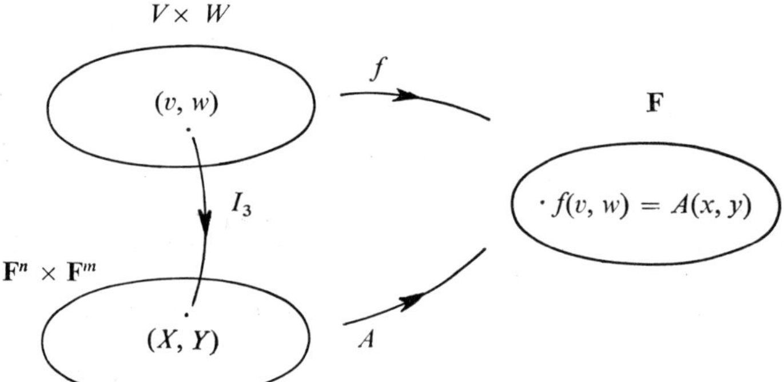


To find a matrix $A$ and an operation of $A$ on the element $(X, Y)$ such that

$$A(X, Y) = f(v, w).$$

We know that

$$f(v, w) = f\left(\sum_{i=1}^{n} a_i v_1, \sum_{j=1}^{m} b_j w_j\right)$$

$$= \sum_{i=1}^{n} \sum_{j=1}^{m} a_i b_j f(v_i, w_j),$$

by the bilinearity of $f$.

Let the scalars $f(v_i, w_j)$ be the elements of a matrix $A$,

$$a_{ij} = f(v_i, w_j).$$

Then $f(v, w) = \sum_{i=1}^{n} \sum_{j=1}^{m} a_i b_j a_{ij}$.

Interpreting $X$ and $Y$ as column vectors this is equivalent to

$$f(v, w) = X^t A Y.$$

We therefore take $A$, the matrix representation of $f$ to be the matrix with elements $a_{ij} = f(v_i, w_j)$ and the operation of $A$ on an element $(X, Y)$ of $\mathbf{F}^n \times \mathbf{F}^m$ to be defined by $A(X, Y) = X^t AY$. $A$ operating thus on $\mathbf{F}^n \times \mathbf{F}^m$ is equivalent to $f$ operating on $V \times W$. $A$ will be a bilinear mapping of $\mathbf{F}^n \times \mathbf{F}^m$ into $\mathbf{F}$.

Note that the elements of $A$ are given by $a_{ij} = f(v_i, w_j)$ and hence depend upon the bases of $V$ and $W$.

*Example 1*

Find the matrix representation of the bilinear form $f$ defined below, relative to the basis $\{(1, 1, 0), (1, 2, 0), (0, 0, 1)\}$ of $\mathbf{R}^3$.

$$f: ((a, b, c), (p, q, r)) \mapsto 2ap + 3bq + cr.$$

We have, using that $a_{ij} = f(v_i, v_j)$,

$a_{11} = f((1, 1, 0), (1, 1, 0)) = 2 + 3 = 5,$

$a_{12} = f((1, 1, 0), (1, 2, 0)) = 2 + 6 = 8,$

$a_{13} = f((1, 1, 0), (0, 0, 1)) = 0,$

$a_{21} = f((1, 2, 0), (1, 1, 0)) = 2 + 6 = 8,$

$a_{22} = f((1, 2, 0), (1, 2, 0)) = 2 + 12 = 14,$

$a_{23} = f((1, 2, 0), (0, 0, 1)) = 0,$

$a_{31} = f((0, 0, 1), (1, 1, 0)) = 0,$

$a_{32} = f((0, 0, 1), (1, 2, 0)) = 0,$

$a_{33} = f((0, 0, 1), (0, 0, 1)) = 1.$

Hence the matrix representation $A$ of $f$ relative to this basis is

$$\begin{pmatrix} 5 & 8 & 0 \\ 8 & 14 & 0 \\ 0 & 0 & 1 \end{pmatrix}$$

We now illustrate the operation of $A$ on $\mathbf{R}^3 \times \mathbf{R}^3$:

$$(a, b, c) = (2a - b)(1, 1, 0) + (b - a)(1, 2, 0) + c(0, 0, 1),$$

and

$$(p, q, r) = (2p - q)(1, 1, 0) + (q - p)(1, 2, 0) + r(0, 0, 1).$$

Hence the representation $X$ of $(a, b, c)$ relative to the given basis is $(2a - b, b - a, c)$ and the representation $Y$ of $(p, q, r)$ is $(2p - q, q - p, r)$. This gives that

$$I_3: ((a, b, c), (p, q, r)) \mapsto ((2a - b, b - a, c), (2p - q, q - p, r)).$$

It can be shown that

$$(2a-b, b-a, c)\begin{pmatrix}5 & 8 & 0\\ 8 & 14 & 0\\ 0 & 0 & 1\end{pmatrix}\begin{pmatrix}2p-q\\ q-p\\ r\end{pmatrix} \begin{aligned}&= 2ap + 3bq + cr\\ &= f((a, b, c), (p, q, r)),\end{aligned}$$

illustrating that $X^tAY = f((a, b, c), (p, q, r))$.

The commutative diagram is:

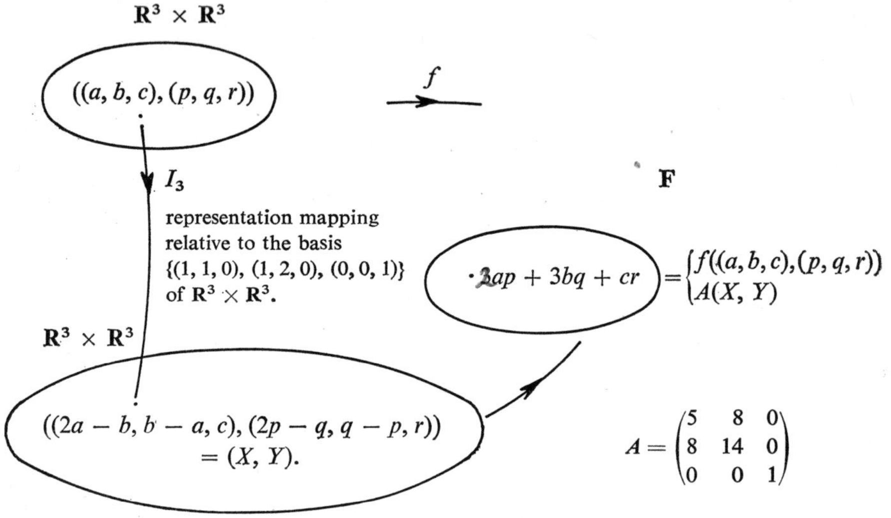


*Theorem 1*

If $B(V, V; F)$ is the set of bilinear forms from $V \times V \rightarrow \mathbf{F}$, where $V$ is a vector space of dimension $n$, then there exists an isomorphism of $B(V, V; \mathbf{F})$ onto the vector space of $n \times n$ matrices with elements in $\mathbf{F}$. There will exist such an isomorphism corresponding to each basis of $V$.

Proof: Let $h$ be the mapping of $B(V, V; \mathbf{F})$ that associates with each element of $B(V, V; \mathbf{F})$ its matrix representation relative to a certain fixed basis $\{v_1, \ldots, v_n\}$ of $V$. We shall show that $h$ is an isomorphism.

First we prove that $h$ is injective. Let $f$ and $g$ be distinct elements of $B(V, V; \mathbf{F})$. Let $h(f) = A$ and $h(g) = B$. Then $A$ has elements $a_{ij}$ given by $a_{ij} = f(v_i, v_j)$ and $B$ has elements $b_{ij}$ given by $b_{ij} = g(v_i, v_j)$. For $v$ and $w$ arbitrary elements of $V$, we know from above that

$$f(v, w) = \sum_{i=1}^{n} \sum_{j=1}^{n} a_i b_j f(v_i, v_j) \quad \text{and} \quad g(v, w) = \sum_{i=1}^{n} \sum_{j=1}^{n} a_i b_j g(v_i, v_j),$$

where $v = \sum_{i=1}^{n} a_i v_i$ and $w = \sum_{j=1}^{n} b_j v_j$. Since $f$ and $g$ are distinct, $f(v, w)$ and $g(v, w)$ must differ on at least one element $(v, w)$ of $V \times V$. Hence $f(v_i, v_j) \neq g(v_i, v_j)$ for at least one pair of base elements. This gives that $A$ and $B$ are distinct, proving that the mapping $h$ is injective.

If $I_3$ is, as previously, the representation isomorphism of $V \times V$ onto $\mathbf{F}^n \times \mathbf{F}^n$, then $h$ is surjective since $AI_3$ defines a bilinear form on $V \times V \to \mathbf{F}$ which has matrix representation $A$.

It remains to prove that the mapping $h$ is linear. For $f$ and $g$, elements of $B(V, V;\mathbf{F})$,

$$\begin{aligned}(f + g)(v, w) &= f(v, w) + g(v, w) \\ &= X^t AY + X^t BY \\ &= X^t(A + B) Y.\end{aligned}$$

Hence $h(f + g) = A + B = h(f) + h(g)$.

Further,

$$\begin{aligned}(af)(v, w) &= a[f(v, w)] \\ &= aX^t AY \\ &= X^t(aA) Y,\end{aligned}$$

implying that

$$h(af) = aA = ah(f).$$

$h$ is thus an isomorphism of $B(V, V; \mathbf{F})$ onto the set of $n \times n$ matrices, corresponding to the basis $\{v_1, \ldots, v_n\}$.

We know that the matrix representation of a bilinear form varies from basis to basis. We now investigate this variation.

*Theorem 2*

Let $A$ and $A'$ be two matrix representations of a bilinear form $f$ of $V \times W \to \mathbf{F}$. Then there exist matrices $N$ and $M$ such that $A = N^t A' M$.

Proof: Let $\{v_1, \ldots, v_n\}$, $\{w_1, \ldots, w_m\}$ and $\{v_1', \ldots, v_n'\}$, $\{w_1', \ldots, w_m'\}$ be the bases of $V$ and $W$ that give representations $A$ and $A'$ of $f$.

Let the representation mappings of $V \times W$ relative to these bases be $I_3$ and $I_3'$ and if $(v, w)$ be an arbitrary element of $V \times W$, let

$$I_3(v, w) = (X, Y) \quad \text{and} \quad I_3'(v, w) = (X', Y').$$

Hence

$$f(v, w) = (X)^t AY = (X')^t A'Y'.$$

But $X$ and $X'$ are representations of the vector $v$ relative to the bases $\{v_1, \ldots, v_n\}$ and $\{v_1', \ldots, v_n'\}$ of $V$. Hence we know that there exists a matrix $N$ such that $X' = NX$. $N$ is in fact that matrix representation of the identity mapping of $V$ relative to these two bases. Similarly there exists a matrix $M$ such that $Y' = MY$.

Therefore

$$\begin{aligned}X^t AY &= (X')^t A'Y', \\ &= (NX)^t A'(MY), \\ &= X^t N^t A' MY.\end{aligned}$$

Since this result is true for all vectors $(X, Y)$ of $\mathbf{F}^n \times \mathbf{F}^m$,

$$A = N^t A' M.$$

In particular, when the bilinear form is of $V \times V \to \mathbf{F}$, this result gives that

$$A = N^t A' N.$$

It is interesting to compare this result with the analogous result for linear mappings discussed in section 6.4.

We now proceed to investigate the characteristics of matrix representations of symmetric bilinear forms.

*Theorem 3*

Let $f$ be a bilinear form of $V \times V \to \mathbf{F}$. If $\{v_1, \ldots, v_n\}$ is a basis for $V$ and $A$ the matrix representation of $f$ relative to this basis, then $A$ is symmetric if and only if $f$ is symmetric.

Proof: Let $v$ and $w$ be arbitrary elements of $V$ and $X$ and $Y$ their representations relative to this basis. Firstly, assume that $A$ is symmetric.

Then

$$\begin{aligned} f(v, w) &= X^t A Y \\ &= (X^t A Y)^t, \text{ since } X^t A Y \text{ is a scalar,} \\ &= (AY)^t X \\ &= Y^t A X \\ &= f(w, v). \end{aligned}$$

Hence $f$ is symmetric.

Conversely, if $f$ is symmetric, then

$$f(v, w) = f(w, v),$$

implying that

$$X^t A Y = Y^t A X.$$

But

$$X^t A Y = (X^t A Y)^t = Y^t (X^t A)^t = Y^t A^t X.$$

Hence

$$Y^t A X = Y^t A^t X$$

for all vectors $X, Y$ of $\mathbf{F}^n$.

This implies that $A = A^t$, hence the matrix $A$ is symmetric.

We saw previously that an inner product on a vector space $V$ is a symmetric bilinear form on $V \times V \to \mathbf{F}$ (section 4.1). Hence an inner product may be represented by a matrix. Any such representation is symmetric. (Interpret your result to exercise 10, section 4.1, in the light of this conclusion.)

The following theorem gives a further property of inner products:

*Theorem 4*

The matrix representation of an inner product relative to an orthogonal basis is a diagonal matrix.

Proof: Let $\{v_1, ..., v_n\}$ be any orthogonal basis for $V$. Then the components of the matrix representation relative to this basis are given by

$$a_{ij} = f(v_i, v_j) = \langle v_i, v_j \rangle = c_i \delta_{ij},$$

for scalars $\{c_1, ..., c_n\}$.

Hence the matrix representation of the inner product relative to this basis is

$$\begin{pmatrix} c_1 & o & \cdots & \\ o & c_2 & & \\ \vdots & & \ddots & \\ & & & c_n \end{pmatrix}.$$

Further, the matrix representation of an inner product relative to an orthonormal basis is the appropriate unit matrix, for here the elements of the matrix representation are given by

$$a_{ij} = f(v_i, v_j) = \langle v_i, v_j \rangle = \delta_{ij}.$$

Such a basis can always be constructed for an inner-product space which has a positive-definite inner product, using the Schmidt orthogonalization process.

Let us now investigate the possibility of representing quadratic forms by matrices. We know that there is a quadratic form associated with every symmetric bilinear form on $V \times V \rightarrow \mathbf{F}$. We now define the matrix representation of a quadratic form and find that it is suitable to take the same matrix as that which represents the bilinear form.

Let $f$ be the symmetric bilinear form on a vector space $V$ of dimension $n$ over a field $\mathbf{F}$ and let $q$ be the associated quadratic form. We shall desire that a matrix representation $A$ of $q$ make the following diagram commutative:

>p. 56

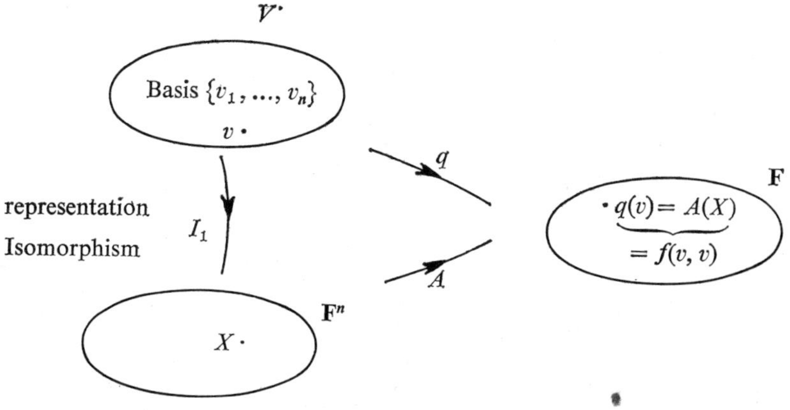

Hence we shall want $A$ to be matrix which under a suitable operation on $\mathbf{F}^n$ makes $A(X) = f(v, v)$. But if we take $A$ to be the matrix representation of $f$ relative to this basis, we have that

$$f(v, v) = X^t AX.$$

Therefore define $A$, the matrix representation of $q$, to be the matrix representation of $f$, and let the operation of $A$ on $\mathbf{F}^n$ be defined by

$$A(X) = X^t AX.$$

The diagram will then be commutative.

Since the matrix representation of any symmetric bilinear form on $V \times V \to \mathbf{F}$ is symmetric, we know that the matrix representation of any quadratic form will also be symmetric.

*Theorem 5*

Let $S(V, V; \mathbf{F})$ be the set of symmetric bilinear forms from $V \times V \to \mathbf{F}$, $Q(V; \mathbf{F})$ be the set of quadratic forms corresponding to $S(V, V; \mathbf{F})$, and if $V$ is of dimension $n$ let $M$ be the set of symmetric $n \times n$ matrices with elements from $\mathbf{F}$. Then the set $Q(V; \mathbf{F})$ forms a vector space and the spaces $S(V, V; \mathbf{F})$, $Q(V; \mathbf{F})$ and $M$ are isomorphic, with an isomorphism corresponding to each basis of $V$.

Proof: Having proved many theorems of this nature, we leave the proof of this theorem to the student!

*Example 2*

The matrix $\begin{pmatrix} 1 & 2 \\ 2 & 3 \end{pmatrix}$ is a symmetric $2 \times 2$ matrix over the field $\mathbf{R}$. It therefore defines both a symmetric bilinear form on $\mathbf{R}^2$ and a quadratic form on $\mathbf{R}^2$. Define these forms.

If $\begin{pmatrix} a \\ b \end{pmatrix}$ and $\begin{pmatrix} p \\ q \end{pmatrix}$ are arbitrary elements of $\mathbf{R}^2$ then the bilinear form $f$ will be defined by

$$f\left(\begin{pmatrix} a \\ b \end{pmatrix}, \begin{pmatrix} p \\ q \end{pmatrix}\right) = (a, b)\begin{pmatrix} 1 & 2 \\ 2 & 3 \end{pmatrix}\begin{pmatrix} p \\ q \end{pmatrix} = (a, b)\begin{pmatrix} p + 2q \\ 2p + 3q \end{pmatrix}$$

$$= ap + 2aq + 2bp + 3bq.$$

The quadratic form will be defined by

$$q\left(\begin{pmatrix} a \\ b \end{pmatrix}\right) = (a, b)\begin{pmatrix} 1 & 2 \\ 2 & 3 \end{pmatrix}\begin{pmatrix} a \\ b \end{pmatrix}$$

$$= a^2 + 4ab + 3b^2.$$

## Exercises

1 Are the following mappings bilinear forms?
a) The mapping of $\mathbf{R} \times \mathbf{R} \to \mathbf{R}$ defined by $(a, b) \mapsto a + b$;
b) The mapping of $\mathbf{R} \times \mathbf{R} \to \mathbf{R}$ defined by $(a, b) \mapsto 3ab$;
c) For $V$ the vector space of continuous real-valued functions on $[0, 1]$, the mapping of $V \times V \to \mathbf{R}$ defined by

$$(f, g) \mapsto \int_0^1 fg\, dt + 4.$$

We know that the dot product is an inner product on the real vector space $\mathbf{R}^3$. Find the matrix representation of the dot product relative to
a) the basis $\{(1, 0, 0), (0, 1, 0), (0, 0, 1)\}$ of $\mathbf{R}^3$;
b) the basis $\{(1, 1, 0), (1, -1, 0), (0, 0, 1)\}$ of $\mathbf{R}^3$.

3 We know from exercise 6, section 4.1 that

$$\langle (a, b, c), (p, q, r) \rangle = 3ap + 3bq + 2cr$$

defines an inner product on the vector space $\mathbf{R}^3$. Find the matrix representation of this inner product relative to the usual basis for $\mathbf{R}^3$.

4 Let $V$ be the real inner-product space of real-valued functions on the interval $[-\pi/2, \pi/2]$ having as basis the set $\{\sin t, \cos t\}$ and inner product defined by $\int_{-\pi/2}^{\pi/2} fg\, dt$. Find the matrix representation of the inner product relative to this basis.

5 Let $V$ the real inner-product space of $2 \times 2$ matrices with inner product defined by $\langle A, B \rangle = \sum_{i=1}^{2} \sum_{j=1}^{2} a_{ij} b_{ij}$, where $A = (a_{ij})$ and $B = (b_{ij})$. Find the matrix representation of this inner product relative to the basis $\left\{ \begin{pmatrix} 1 & 0 \\ 0 & 0 \end{pmatrix}, \begin{pmatrix} 0 & 1 \\ 0 & 0 \end{pmatrix}, \begin{pmatrix} 0 & 0 \\ 1 & 0 \end{pmatrix}, \begin{pmatrix} 0 & 0 \\ 0 & 1 \end{pmatrix} \right\}$ of $V$.

6 Let $V$ be the vector space of all polynomials over $\mathbf{R}$ of degree $\leqq 1$ generated by the basis $\{1, t\}$. A bilinear form $\mathbf{F}$ is defined on $V$ by

$$\mathbf{F}(f, g) = \int_0^1 fg\, dt.$$

Find the matrix representation of $F$ relative to this basis.

7 Let $V$ and $W$ be vector spaces of dimensions $n$ and $m$ respectively over a common field of scalars $\mathbf{F}$. Let $I_1$ and $I_2$ be representation isomorphisms of $V$ onto $\mathbf{F}^n$ and $W$ onto $\mathbf{F}^m$ with $I_1(v) = X$ and $I_2(w) = Y$, for arbitrary vectors $v$ and $w$ of $V$ and $W$ respectively. In this section we stated that $I_3$ defined by $I_3(v, w) = (X, Y)$ was a structure preserving mapping of $V \times W$ onto $\mathbf{F}^n \times \mathbf{F}^m$. Discuss the meaning of this statement and prove its validity.

8 Commutative diagrams are used extensively io mathematics. Let $A$ and $B$ be sets having binary operations $\cdot$ and $*$ respectively. Analyze the following commutative diagram

$$\begin{array}{ccc} A \times A & \xrightarrow{f \times f} & B \times B \\ \cdot \big\downarrow & & \big\downarrow * \\ A & \xrightarrow{f} & B \end{array}$$

where $f: A \to B$ and $f \times f: A \times A \to B \times B$ is defined by $f \times f(a_1, a_2) = (b_1, b_2)$, where $f(a_1) = b_1$ and $f(a_2) = b_2$. Prove that a mapping $f$ that satisfies the conditions of this diagram is a hormorphism.

## CHAPTER 7

# Tensor Algebra

## 7.1 Tensors

We now come to a discussion of the algebra of tensors. Tensors are the basic tools of much of differential geometry. They are also used by applied mathematicians and theoretical physicists; Einstein formulated his theory of general relativity in terms of tensors. In differential geometry and theoretical physics, this theory is usually interpreted for vector spaces having as base vectors tangent vectors to certain curves. In this framework one usually develops a differential theory of tensors. In this chapter we give the algebraic foundation of tensors.

Let $V$ be a finite-dimensional vector space over a field $\mathbf{F}$ and let $V^*$ be its dual. A $p$-linear form of $V^p$ into $\mathbf{F}$ is called a *covariant tensor of order* $p$; a $q$-linear form of $(V^*)^q$ into $\mathbf{F}$ is called a *contravariant tensor of order* $q$ and a $p+q$-linear form of $V^p \times (V^*)^q$ into $\mathbf{F}$ is called a *mixed tensor* of covariant order $p$ and of contravariant order $q$.

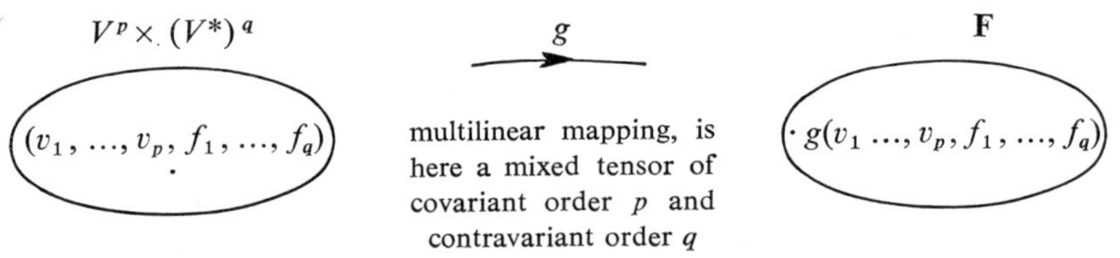


Two tensors are said to be of the *same kind* if they are of the same contravariant order and are of the same covariant order.

We know that, relative to a vector space $V$, all tensors of the same kind, being multilinear mappings on the same space, form a vector space. Operations of addition and scalar multiplication are defined in the usual way for a set of multilinear mappings.

Since $V^*$ is the space of functionals on $V$, that is the space of 1-linear forms on $V$, this is the space of covariant tensors of order one. Such tensors are called *covariant vectors*. We also know that $V$ can be interpreted as the space of functionals on $V^*$. Hence the elements of $V$ are contravariant tensors of order one. These elements are called *contravariant vectors*.

*Example 1*

Let $V$ be a vector space and $g$ an inner product on $V$. Then $g$ is a covariant tensor of order two for it is a bilinear form of $V \times V$ into the scalar field.

We have seen how bilinear forms can be represented by matrices. Those bilinear forms involving one vector space and its dual are tensors of order two. Such tensors can therefore be represented by matrices. We will now see how any tensor can be represented by a set of components—an extension of the matrix representation concept for second order tensors.

Let $V$ be a vector space of dimension $n$ over a field $\mathbf{F}$ and $t$ a mixed tensor of covariant order $p$ and of contravariant order $q$. Then a natural structure-preserving mapping $I_1$ exists of $(V)^p \times (V^*)^q$ onto $(\mathbf{F}^n)^p \times (\mathbf{F}^n)^q$ corresponding to every basis of $V$ (the dual basis is taken for $V^*$). We shall define in terms of a set of scalars $T$ a mapping of $(\mathbf{F}^n)^p \times (\mathbf{F}^n)^q$ into $\mathbf{F}$ that has the same properties relative to $(\mathbf{F}^n)^p \times (\mathbf{F}^n)^q$ as $t$ has relative to $(V)^p \times (V^*)^q$. The set $T$ will then be called the representation of the tensor $t$ relative to this basis for $V$.

The situation can be represented by the following commutative diagram:

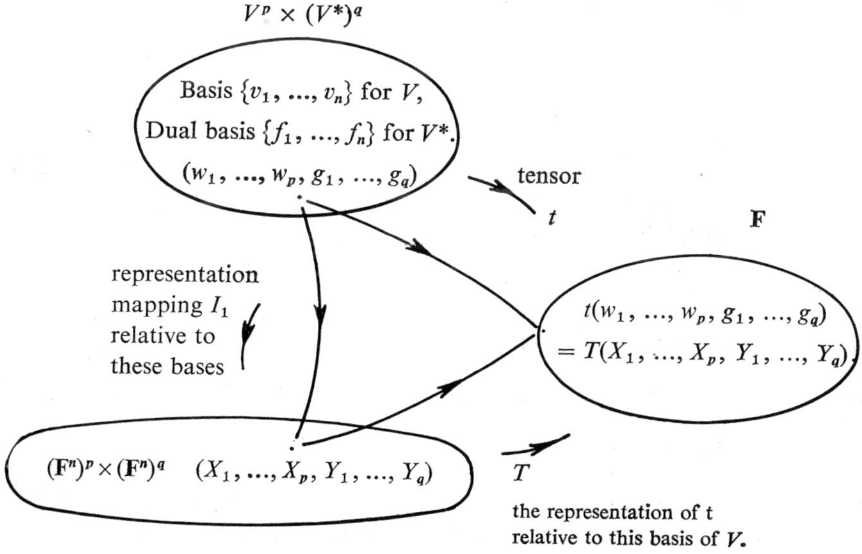


The procedures that we use to find the set $T$ and the manner in which this set operates on $(\mathbf{F}^n)^p \times (\mathbf{F}^n)^q$ are the same as in the matrix representation of a bilinear form.

Since the sets $\{v_1, \ldots, v_n\}$, $\{f_1, \ldots, f_n\}$ are bases for $V$ and $V^*$ then an arbitrary element $(w_1, \ldots, w_p, g_1, \ldots, g_q)$ of $V^p \times (V^*)^q$ can be written $\left(\sum_a r_a v_a, \ldots, \sum_b s_b v_b, \sum_c q_c f_c, \ldots, \sum_d p_d f_d\right)$.

p of these q of these

Hence

$$t(w_1, \ldots, w_p, g_1, \ldots, g_q) = t\left(\sum_a r_a v_a, \ldots, \sum_b s_b v_b, \sum_c q_c f_c, \ldots, \sum_d p_d f_d\right).$$

Using the multilinearity of $t$ this becomes

$$t(w_1, \ldots, w_p, g_1, \ldots, g_q) = \sum_a \cdots \sum_d r_a \ldots s_b q_c \ldots p_d\, t(v_a, \ldots, v_b, f_c, \ldots, f_d).$$

Let the scalars $t(v_a, \ldots, v_b, f_c, \ldots, f_d)$, $1 \leqq a \leqq n, \ldots, 1 \leqq d \leqq n$, be the elements of the set $T$ that gives the representation of $t$ relative to this basis of $V$. Hence, as in the case of a bilinear form, the representation of a tensor is given by its values on the base vectors of the vector space and its dual. Label these elements

$$t(v_a, \ldots, v_b, f_c, \ldots, f_d) = T^{c\ldots d}_{a\ldots b}, \quad 1 \leqq a \leqq n, \ldots, 1 \leqq d \leqq n.$$

The set $\{T^{c\ldots d}_{a\ldots b}\}$, $1 \leqq a \leqq n, \ldots, 1 \leqq d \leqq n$ is defined to be the representation of $t$ relative to this basis for $V$.

The operation of $T$ on an element of $(\mathbf{F}^n)^p \times (\mathbf{F}^n)^q$ is defined by

$$\begin{aligned} T(X_1, \ldots, X_p, Y_1, \ldots, Y_q) &= t(w_1, \ldots, w_p, g_1, \ldots, g_q) \\ &= \sum_{a=1}^{n} \cdots \sum_{d=1}^{n} r_a \ldots s_b q_c \ldots p_d T^{c\ldots d}_{a\ldots b}, \end{aligned}$$

where $X_1 = (r_1, \ldots, r_n), \ldots, Y_q = (p_1, \ldots, p_n)$.

Here $X_i$ represents the vector $w_i$, which being an element of $V$ is a contravariant vector and $Y_j$ represents $f_j$, which being an element of $V^*$ is a covariant vector. The usual notation for the components of such representations in tensor analysis is

$$X_i = \left(\underset{i}{X^1}, \underset{i}{X^2}, \ldots, \underset{i}{X^n}\right) \quad \text{and} \quad Y_j = \left(\underset{j}{Y_1}, \underset{j}{Y_2}, \ldots, \underset{j}{Y_n}\right),$$

We can represent $X_i$ by an arbitrary component $\underset{i}{X^a}$ and $Y_j$ by an arbitrary component $\underset{j}{Y_a}$. This concept is similar to representing a matrix $A$ by an arbitrary element $a_{ij}$. Superscripts are used for the components of a contravariant vector and subscripts for the components of a covariant vector. Using this new notation,

$$T(X_1, \ldots, X_p, Y_1, \ldots, Y_q) = \sum_{a=1}^{n} \cdots \sum_{d=1}^{n} \underset{1}{X^a} \ldots \underset{p}{X^b}\, \underset{1}{Y_c} \ldots \underset{q}{Y_d}\, T^{c\ldots d}_{a\ldots b}$$

A *summation convention* is a further simplification used in tensor analysis: when a suffix is repeated, summation is understood on that suffix, over the relevant range. Using this convention the operation of $T$ on $(\mathbf{F}^n)^p$

$\times\ (\mathbf{F}^n)^q$ is defined by

$$T(X_1, \ldots, X_p, Y_1, \ldots, Y_q) = T^{c\ldots d}_{a\ldots b} \underset{1}{X^a} \cdots \underset{p}{X^b}\, \underset{1}{Y_c} \ldots \underset{q}{Y_d}. \tag{1}$$

The usefulness of employing superscripts for contravariant vectors and subscripts for covariant vectors can now be seen—repeated indices, implying summation over those indices occur one at the superscript level and the other at the subscript level. This is generally the case in tensor analysis.

A further interpretation of the components of this representation is of interest. This makes use of the concept of the *tensor product*, which we now define.

Let $\{V_i\}$, $i = 1, \ldots, p$ be a set of finite-dimensional vector spaces over a common field of scalars $\mathbf{F}$. Let $f_i \in V_i^*$, for all $i = 1, \ldots, p$. Then there exists a $p$-linear form $f_1 \otimes \cdots \otimes f_p$ of $V_1 \times \cdots \times V_p \to \mathbf{F}$ defined by

$$f_1 \otimes \cdots \otimes f_p(v_1, \ldots, v_p) = f_1(v_1)\, f_2(v_2) \ldots f_p(v_p),$$

for every $(v_1, \ldots, v_p)$ element of $V_1 \times \cdots \times V_p$. The proof that this mapping is $p$-linear is left as an exercise. This mapping is called the tensor product of $f_1, \ldots, f_p$.

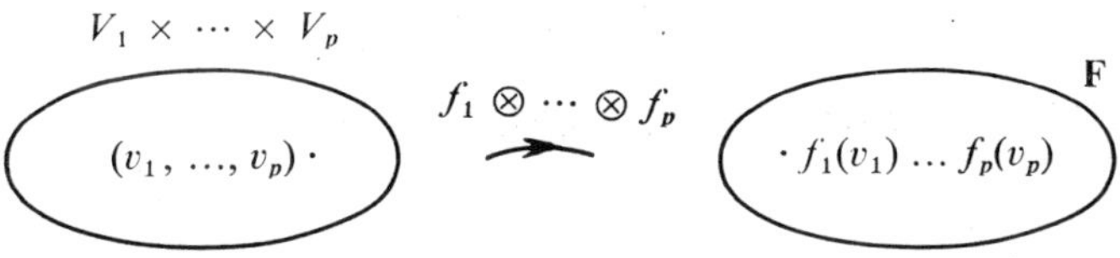


We now return to the discussion of the representation of the tensor $t$.

$$t(w_1, \ldots, w_p, g_1, \ldots, g_q) = T^{c\cdots d}_{a\cdots b} \underset{1}{X^a} \ldots \underset{p}{X^b}\, \underset{1}{Y_c} \ldots \underset{q}{Y_d}.$$

Here $\underset{i}{X^l}$ is the $l$th component of $w_i$ relative to the basis $v_1, \ldots, v_n$. Hence $\underset{i}{X^l} = f_l(w_i)$.

Similarly, interpreting $V$ as $V^{**}$, $\underset{j}{Y_m}$ is the $m$th component of $g_j$ relative to the basis $f_1, \ldots, f_n$. Hence $\underset{j}{Y_m} = v_m(g_j)$.

This leads to

$$\begin{aligned} t(w_1, \ldots, w_p, g_1, \ldots, g_q) &= T^{c\cdots d}_{a\cdots b} f_a(w_1) \ldots f_b(w_p)\, v_c(g_1) \ldots v_d(g_q) \\ &= T^{c\cdots d}_{a\cdots b} f_a \otimes \ldots f_b \otimes v_c \ldots \otimes v_d(w_1, \ldots, w_p, g_1, \ldots, g_q). \end{aligned}$$

Hence

$$t = T^{c\cdots d}_{a\cdots b} f_a \otimes \ldots f_b \otimes v_c \ldots \otimes v_d.$$

Thus, the set $\{f_a \otimes \cdots f_b \otimes v_c \cdots \otimes v_d\}$ spans the space $M(V^p, (V^*)^q; \mathbf{F})$ of multilinear forms on $V^p \times (V^*)^q$. It can be shown (left as an exercise), that this set is also linearly independent and hence forms a basis for this space. The set $T$ is thus the representation of the tensor $t$ relative to this basis.[1]

*Example*

Let $\mathbf{R}^4$ be an inner product space with inner product $g$ defined by

$$g((x_1, x_2, x_3, x_4), (y_1, y_2, y_3, y_4)) = x_1y_1 + x_2y_2 + x_3y_3 - x_4y_4.$$

Hence $\mathbf{R}^4$ is the inner product space of Special Relativity called *Minkowski Space* discussed in exercise 8, section 4.2. $g$ is thus a covariant tensor of order 2. Find its representation $G$ relative to the usual basis of $\mathbf{R}^4$.

The commutative diagram is

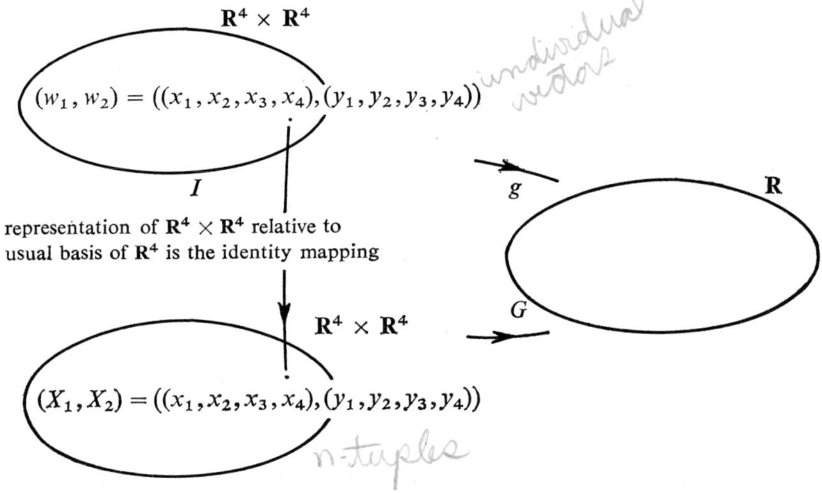


Let the elements of $G$ be $G_{ab}$, $1 \leqq a \leqq 4$, $1 \leqq b \leqq 4$. We label the elements of $G$ thus since $g$ is a covariant tensor of order 2. These components are given by

$$G_{ab} = g(v_a, v_b),$$

where $\{v_1, v_2, v_3, v_4\}$ is the usual basis for $\mathbf{R}^4$. This gives $G_{11} = g((1, 0, 0, 0), (1, 0, 0, 0)) = 1$, $G_{12} = g((1, 0, 0, 0), (0, 1, 0, 0)) = 0$, $G_{22} = G_{33} = 1$, $G_{44} = -1$. All the other elements are identically zero. Since we have only two indices here it is convenient to write $G$ as the matrix

$$\begin{pmatrix} 1 & 0 & 0 & 0 \\ 0 & 1 & 0 & 0 \\ 0 & 0 & 1 & 0 \\ 0 & 0 & 0 & -1 \end{pmatrix}$$

[1] Remember that the summation convention holds on repeated indices here. The rule that repeated indices appear once as subscripts and once as superscripts could have been adhered to if we had represented the dual basis of $V^*$ by $\{f^1, \ldots, f^n\}$ instead of as $\{f_1, \ldots, f_n\}$. This is the way it is usually written in tensor analysis. However in this introduction this notation would be confusing since one would tend to confuse vectors and components of vectors.

In general, of course, it is not possible to write down the components of the representation of a tensor as a matrix.

Note that this matrix would also be the matrix representation of $g$ relative to the usual basis. This is to be expected since we defined the concept of a tensor representation as a generalization of the concept of matrix representation for a bilinear form.

For $(w_1, w_2)$ an arbitrary element of $\mathbf{R}^4 \times \mathbf{R}^4$ with representation $(X_1, X_2)$, we have that the effect of $g$ on $\mathbf{R}^4 \times \mathbf{R}^4$ can be represented by $G$ on $\mathbf{R}^4 \times \mathbf{R}^4$ in the following manner:

$$g(w_1, w_2) = G_{ab}\underset{1}{X^a}\underset{2}{X^b}.$$

Using the known components of $G$ this reduces to

$$\begin{aligned} g(w_1, w_2) &= G_{11}\underset{1}{X^1}\underset{2}{X^1} + G_{22}\underset{1}{X^2}\underset{2}{X^2} + G_{33}\underset{1}{X^3}\underset{2}{X^3} - G_{44}\underset{1}{X^4}\underset{2}{X^4} \\ &= x_1y_1 + x_2y_2 + x_3y_3 - x_4y_4, \end{aligned}$$

verifying the representation.

### Exercises

1 Find the tensor representation of the inner product of the example of this section relative to the basis a) (1, 1, 0, 0), (1, −1, 0, 0), (0, 0, 1, 0), (0, 0, 0, 1) of $\mathbf{R}^4$; b) (1, 2, 0, 0), (−2, 1, 1, 0), (0, 3, 2, 0), (0, 0, 0, 1) of $\mathbf{R}^4$.

2 Let $V$ be the vector space $\mathbf{R}^4$ with inner product defined to be the dot product. The inner product is a covariant tensor of order two. Find its representation a) relative to the usual basis of $\mathbf{R}^4$; b) relative to the basis {(1, 0, 1, 0), (−1, 0, 1, 0), (0, 1, 0, 1), (0, −1, 0, 1)} of $\mathbf{R}^4$.

3 Let $t$ be a multilinear mapping of $V^3 \times (V^*)^4$ into $\mathbf{F}$. What type of tensor is $t$? How many elements are there in a representation $T$ of $t$ and how are they labeled?

4 Let $t$ be a multilinear mapping of $V^4 \times V^*$ into $\mathbf{F}$. What type of tensor is $t$? How many elements are there in a representation of $T$ of $t$ and how are they labeled?

5 Let $g$ be a tensor with representation $G = \{G^{abc}_{de} : 1 \leqq a \leqq 2, \ldots, 1 \leqq e \leqq 2\}$. What kind of tensor is $g$? What is the dimension of the vector space which generates the domain of $g$?

6 Let $f$ be a tensor with representation $F = \{F^a_b : 1 \leqq a \leqq 3, 1 \leqq b \leqq 3\}$. What kind of tensor is $f$? What is the dimension of the vector space which generates the domain of $f$?

7 The following mapping $f$ is a linear mapping of $\mathbf{R}^2$ into $\mathbf{R}$:

$$f(x, y) = x.$$

It may thus be interpreted as a covariant vector. Find the representation of this covariant vector a) using the usual basis for $\mathbf{R}^2$; b) using the basis {(1, 1), (1, 2)} for $\mathbf{R}^2$.

8 Let $\{V_i\}$, $i = 1, \ldots, p$, be a set of finite dimensional vector spaces of a common field of scalars $\mathbf{F}$. Let $f_i \in V_i^*$ for all $i = 1, \ldots, p$. Prove that $f_1 \otimes \cdots \otimes f_p$ defined as a mapping of $V_1 \times \cdots \times V_p \to \mathbf{F}$ by

$$f_1 \otimes \cdots \otimes f_p(v_1, \ldots, v_p) = f_1(v_1) \cdots f_p(v_p),$$

is $p$-linear.

9 Let $V$ be a vector space and $V^*$ its dual. If $f_1$ and $f_2$ are arbitrary elements of $V^*$ show that $f_1 \otimes f_2 \neq f_2 \otimes f_1$ showing that the tensor product is not in general commutative.

10 Let $V$ a vector space over a field **F**. a) Describe the tensor that has representation $\{\delta^a_b : 1 \leq a \leq n, 1 \leq b \leq n\}$, relative to a given basis for $V$. b) Describe the tensor that has representation $\{\delta_{ab} : 1 \leq a \leq n, 1 \leq b \leq n\}$, relative to the given basis.

## 7.2 The Reisz Representation Theorem in Tensor Form

Let us now interpret the Reisz representation theorem in the language of tensors. Let $V$ be a vector space of dimension $n$ over a field **F** having an inner product $g$ with dual space $V^*$. Let $\{v_1, \ldots, v_n\}$ be a basis for $V$ and $\{f_1, \ldots, f_n\}$ be the dual basis. Then we know that there exists a natural isomorphism $h$ of $V$ onto $V^*$ defined by $v \leftrightarrow f_v$, where $f_v(w) = g(w, v)$. Let $I_1$ be the representation isomorphism of $V$ onto $\mathbf{F}^n$ corresponding to this basis and $I_2$ be the isomorphism of $V^*$ onto $\mathbf{F}^n$ corresponding to the dual basis. Then $h$ will have a representation $H$ of $\mathbf{F}^n$ onto $\mathbf{F}^n$ that makes the following diagram commutative. Here we wish to define $H$.

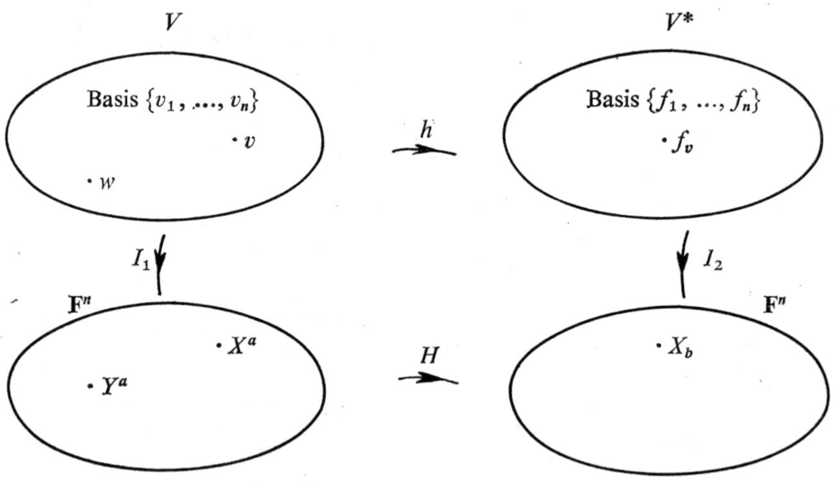


Using tensor notation, let the representation of $v$ be $X^a$, w be $Y^a$ and $f_v$ be $X_b$ where $f_v$ is the image of $v$ under $h$ and $w$ is in arbitrary element of $V$. There should be no confusion in using the same letter $X$ with a superscript to denote the element of $V$ and with a subscript to denote its image under $H$. $X_b$ being the image of $f_v$ under $I_2$ is defined by

$$f_v = \sum_{i=1}^{n} X_i f_i.$$

(It is the representation of $f_v$ relative to the dual basis.) This gives

$$f_v(v_j) = \sum_{i=1}^{n} X_i f_i(v_j) = X_j,$$

implying that

$$g(v_j, v) = X_j.$$

We now use the representation $G_{ab}$ of $g$ to write this result

$$X_j = G_{pq} \underset{j}{X^p} X^q$$

where $\underset{j}{X^p}$ is the image of $v_j$ under $I_1$. But $\underset{j}{X^p}$, being the representation of the base vector $v_j$, has components (0, 0, …, 1, …, 0) with 1 in the $j$th slot.

Hence

$$X_j = G_{jq} X^q.$$

The representation $H$ of the isomorphism $h$ is therefore defined by

$$H: X^a \mapsto X_b,$$

where $X_b = G_{bc} X^c$. In tensor analysis this mapping is called *lowering the index* using the inner product. Hence the natural isomorphism between an inner-product space and its dual is defined in tensor language by lowering the index, using the tensor representation of the inner product. Note that $H$ can also be interpreted as a matrix, the matrix representation of $h$ relative to these bases. This matrix will be $(G_{ab})$. The student is asked to verify this in the exercises.

Further, since $f_v$ is a functional on $V$ and $X_b$ is a representation of $f_v$, $X_b$ is a functional on $\mathbf{F}^n$. It is illustrative to examine the operation of this functional $X_b$ on $\mathbf{F}^n$. We have that

$$f_v(w) = g(w, v) = g(v, w),$$

for an arbitrary $w \in V$. If the representation of $w$ is $Y^q$ this implies that

$$X_b(Y^a) = G_{pq} X^p Y^q,$$

giving

$$X_b(Y^a) = X_q Y^q,$$

remembering summation on the repeated index. This operation is called *contraction* in tensor analysis.

*Example*

Let $\mathbf{R}^4$ be the Minkowski Space discussed in the previous section. Discuss the representation of the natural isomorphism between $\mathbf{R}^4$ and $(\mathbf{R}^4)^*$ generated by the inner product, using the usual basis for $\mathbf{R}^4$ and the dual basis for $(\mathbf{R}^4)^*$.

The commutative diagram is

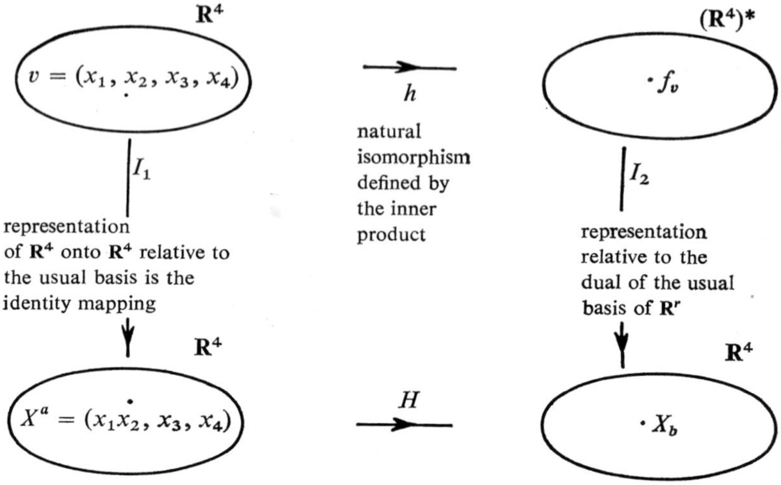


We have from the previous section that $G = (G_{ab})$, the representation of the inner product, is the matrix

$$\begin{pmatrix} 1 & 0 & 0 & 0 \\ 0 & 1 & 0 & 0 \\ 0 & 0 & 1 & 0 \\ 0 & 0 & 0 & -1 \end{pmatrix}.$$

Hence $H\colon X^a \mapsto X_b$, where $X_b = G_{bc}X^c$.

This implies that $X_b$ is given by

$$X_1 = G_{1c}X^c = x_1, \quad X_2 = G_{2c}X^c = x_2, \quad X_3 = G_{3c}X^c = x_3$$

and

$$X_4 = G_{4c}X^c = -x_4.$$

Therefore $X_b$ is the vector $(x_1, x_2, x_3, -x_4)$.

$H$ is defined by

$$H\colon (x_1, x_2, x_3, x_4) \mapsto (x_1, x_2, x_3, -x_4).$$

## Exercises

1 Let $V$ be the vector space $\mathbf{R}^4$ with inner product defined to be the dot product. Let $h$ be the isomorphism of $V$ onto $V^*$ defined by the dot product and $H$ be the representation of $h$ relative to the usual basis of $\mathbf{R}^4$. What is the image of the vector $(1, 2, 3, 4)$ of $\mathbf{R}^4$ under $H$. What is the image of an arbitrary element $X^a$ of $\mathbf{R}^4$ under $H$, that is, what is $X_a$ here? Illustrate with a commutative diagram. (The representation of the dot product relative to this basis has been found in question 2 of the previous exercises.)

2 With $V$ the same inner product space as in question 1, find the representation $H'$ of the natural isomorphism generated by the dot product relative to the basis $\{(1,0,1,0), (-1,0,1,0), (0,1,0,1), (0,-1,0,1)\}$ of $\mathbf{R}^4$. What is the image of an arbitrary element $Y^a$ of $\mathbf{R}^4$ under $H'$, that is, what is $Y_a$ here? Illustrate with a commutative diagram. (The representation of the dot product relative to this basis has been found in question 2 of the previous exercises.)

3 Let $V$ be the vector space with inner product $g$ discussed in this section, $h$ the isomorphism of $V$ onto $V^*$ generated by the inner product and $H$ the representation of $h$ relative to the given basis of $V$ and its dual in $V^*$. Discuss the representation of the inverse of $h$, $h^{-1}$ by a mapping $H^{-1}$ of $X_b$ into $X^a$. Show that there exists a set of scalars $\{G^{ab}: 1 \leqq a \leqq n, 1 \leqq b \leqq n\}$ such that

$$H^{-1}: X_b \mapsto X^a,$$

where $X^a = G^{ab}X_b$. Prove that $G^{ab}G_{bc} = \delta^a_c$.

4 a) Prove that the representation $H$ of the isomorphism of this section can be interpreted as a matrix, the matrix $(G_{ab})$.

b) Prove that the representation $H^{-1}$ of the isomorphism $h^{-1}$ of $V^*$ onto $V$ in question 3, can be interpreted as a matrix, the matrix $(G^{ab})$.

c) Prove that the result $G^{ab}G_{bc} = \delta^a_c$ is equivalent in matrix notation to $(G^{ab})(G_{ab}) = I$, where $I$ is the appropriate unit matrix, showing that $(G^{ab})$ is the inverse of the matrix $(G_{ab})$.

d) Satisfy yourselves that $h$, the natural isomorphism of $V$ onto $V^*$ generated by the inner product, is the mapping of $V$ onto $V^*$ that has as its matrix representation relative to any basis of $V$ and its dual in $V^*$, the matrix representation of the inner product relative to these bases.

## 7.3 The Transformation Law for Tensors

We have seen how representations of tensors exist relative to bases. The next topic to be discussed is the variation of this representation relative to different bases. We have already discussed the variations of the representations of linear mappings and bilinear forms relative to different bases. The procedure used to find the relation between two representations of a tensor will be similar. The rule relating two arbitrary representations of the same tensor is called the *transformation law* for that tensor.

Let $t$ be the tensor considered previously with $\{v_1, \ldots, v_n\}$, $\{v_1', \ldots, v_n'\}$ a pair of bases for $V$ and $\{f_1, \ldots, f_n\}$ and $\{f_1', \ldots, f_n'\}$ the corresponding dual bases in $V^*$. Let $T$ and $T'$ be the sets that represent $t$ relative to these bases, with elements $T^{c\ldots d}_{a\ldots b}$, $1 \leqq a \leqq n, \ldots, 1 \leqq d \leqq n$, and $T'^{k\ldots m}_{r\ldots s}$, $1 \leqq r \leqq n, \ldots, 1 \leqq m \leqq n$ respectively. For $(w_1, \ldots, w_p, g_1, \ldots, g_q)$ an arbitrary element of $V^p \times (V^*)^q$ with representations $(X_1, \ldots, X_p, Y_1, \ldots, Y_q)$ and $(X_1', \ldots, X_p', Y_1', \ldots, Y_q')$ relative to these bases, we have that

$$t(w_1 \ldots w_p, g_1 \ldots g_q) = T^{c\cdots d}_{a\cdots b} \underset{1}{X^a} \ldots \underset{p}{X^b}\, \underset{1}{Y_c} \ldots \underset{q}{Y_d}$$

and

$$t(w_1 \ldots w_p, g_1 \ldots g_q) = T'^{k\cdots m}_{r\cdots s} \underset{1}{X'^r} \ldots \underset{p}{X'^s}\, \underset{1}{Y'_k} \ldots \underset{q}{Y'_m},$$

giving

$$T^{c\cdots d}_{a\cdots b}\underset{1}{X}^a\cdots\underset{p}{X}^b\underset{1}{Y}_c\cdots\underset{q}{Y}_d = T'^{k\cdots m}_{t\cdots s}\underset{1}{X}'^r\cdots\underset{p}{X}'^s\underset{1}{Y}'\cdots\underset{q}{Y}'_m.$$

But $\underset{i}{X}^a$ and $\underset{i}{X}'^b$ are both representations of the same vector $w_i$, relative to different bases $\{v_1, \ldots, v_n\}$ and $\{v_1', \ldots, v_n'\}$. We therefore know that they are related through a matrix $N$ which is in fact the matrix representation of the identity mapping relative to these bases, $(\underset{i}{X}'^a) = N(\underset{i}{X}^b)$. Let the elements of $N$ be written, $x_p^q$, wehre $q$ indicates row and $p$ column. Then the above can be written $\underset{i}{X}'^a = x_b^a\underset{i}{X}^b$. This gives the arbitrary component of the vector $X_i'$.

Similarly the covariant representations $\underset{i}{Y}_a$ and $\underset{i}{Y}_b'$ are related. These are representations of the vector $g_i$ relative to the bases $\{f_1, \ldots, f_n\}$ and $\{f_1', \ldots, f_n'\}$. Hence if $M$ is the matrix representation of the identity mapping of $V^*$ relative to these bases $(\underset{i}{Y}_a') = M(\underset{i}{Y}_b)$. Again let $y_p^q$ be the elements of $M$, now with $p$ indicating row and $q$ indicating column. Then we have that $\underset{i}{Y}_a' = y_a^b\underset{i}{Y}_b$.

On substituting for $\underset{i}{X}'^p$ and $\underset{i}{Y}_q'$ we therefore get that

$$T^{c\cdots d}_{a\cdots b}\underset{1}{X}^a\ldots\underset{p}{X}^b\underset{1}{Y}_c\ldots\underset{q}{Y}_d = T'^{k\cdots m}_{r\cdots s}(x_a^r\underset{1}{X}^a)\ldots(x_b^s\underset{p}{X}^b)(y_k^c\underset{1}{Y}_c)\ldots(y_m^d\underset{q}{Y}_d)$$

giving

$$(T^{c\cdots d}_{a\cdots b} - T'^{k\cdots m}_{r\cdots s}x_a^r\ldots x_b^s y_k^c\ldots y_m^d)(\underset{1}{X}^a\ldots\underset{p}{X}^b\underset{1}{Y}_c\ldots\underset{q}{Y}_d) = O.$$

Since the vectors $w_1, \ldots, w_p, g_1, \ldots, g_q$ are arbitrary elements of the appropriate spaces the vectors $\underset{1}{X}^a, \ldots, \underset{p}{X}^b, \underset{1}{Y}_c, \ldots, \underset{q}{Y}_d$ are arbitrary elements of $\mathbf{F}^n$.

Hence

$$T^{c\cdots d}_{a\cdots b} - T'^{k\cdots m}_{r\cdots s}x_a^r\ldots x_b^s y_k^c\ldots y_m^d = O,$$

implying that

$$T^{c\cdots d}_{a\cdots b} = T'^{k\cdots m}_{r\cdots s}x_a^r\ldots x_b^s y_k^c\ldots y_m^d.$$

This gives the rule that relates different representations of the same tensor. It is the transformation law for tensors.

## Exercises

1 Let $\mathbf{R}^4$ be Minkowski space. Representations of the inner product, which is a covariant tensor of order two relative to two bases, have been found in exercise 1 of section 7.1. Prove that these two representations satisfy the tensor law of transformation. Find the representation relative to the basis $\{(1, 2, 3, 0), (1, 3, 2, 0), (2, 3, 1, 0), (0, 0, 0, 1)\}$ using the tensor law of transformation and one of these representations.

2 The representations of the covariant vector $f$ which is the linear mapping of $\mathbf{R}^2$ into $\mathbf{R}$ defined by $f(x, y) = x$ relative to two bases, have been found in exercise 7 of section 7.1. Prove that these two representations satisfy the tensor law of transformation. Find the representation of this covariant vector relative to the basis $\{(1, -1), (2, 1)\}$ for $\mathbf{R}^2$, using the tensor law of transformation and one of these representations.

CHAPTER 8

# Spectral Theory of Operators

## 8.1 Eigenvectors and Eigenvalues of Operators

In this section we generalize the concepts of eigenvalues and eigenvectors that were previously introduced for matrices. The theory developed also gives us further insight into these concepts in the case of matrices.

Let $V$ be a vector space over a field $\mathbf{F}$ and let $L$ be an operator on $V$. That is, $L$ is a linear mapping of $V$ into itself. A non-zero element $v$ of $V$ is said to be an eigenvector of $L$ if there exists a scalar $\lambda$ such that

$$Lv = \lambda v.$$ [1]

If there exists such a $\lambda$ it is uniquely determined for the eigenvector, for suppose there exist two such $\lambda$'s, $\lambda_1$ and $\lambda_2$. Then $\lambda_1 v = Lv = \lambda_2 v$, implying that $\lambda_1 = \lambda_2$. $\lambda$ is called the *eigenvalue* corresponding to the eigenvector $v$ and the set of all eigenvalues is called the *spectrum of* the operator.

If $V_\lambda$ is the subset of $V$ consisting of all eigenvectors of $L$ corresponding to the eigenvalue $\lambda$, and the zero vector, then it can be proved that $V_\lambda$ is a subspace of $V$. The proof is similar to the proof of this result for matrices (theorem 1, section 5.10). The subspace $V_\lambda$ is called the *eigenspace* of $L$ corresponding to the eigenvalue $\lambda$.

Intuitively an eigenvector of an operator $L$ is a vector that remains unchanged or reversed in direction when $L$ operates on the vector space. If the corresponding eigenvalue is 1 then it is unchanged in magnitude also and is then said to be invariant under $L$.

One importance of eigenvectors is that if there are sufficient to form a basis for the vector space they can be used to give a diagonal matrix representation of the operator $L$. We shall discuss this property later in this section. First we give a method for determining the eigenvectors and corresponding eigenvalues of an operator. Use will be made of a representation space.

*Theorem 1*

Let $\{v_1, \ldots, v_n\}$ be an arbitrary basis for $V$ and let $A$ be the matrix representation of $L$ relative to this basis. Let $I_1$ be the isomorphism of $V$ onto $\mathbf{F}^n$ generated by this basis. Then the eigenvectors of $L$ in $V$ are in

[1] In this chapter we omit the bracket around the domain element. If $L\colon A \to B$ and $a$ is an element of $A$ with image $b$ under $L$ we simply write $La = b$.

one-to-one correspondence under the isomorphism $I_1$ to the eigenvectors of $A$ in $\mathbf{F}^n$. The eigenvalues of corresponding eigenvectors are identical.

This is a result that we would expect since mathematically the structure $L$: $V \to V$ is identical to $A: \mathbf{F}^n \to \mathbf{F}^n$ under the correspondence $I_1$.

The theorem can be illustrated diagramatically thus:

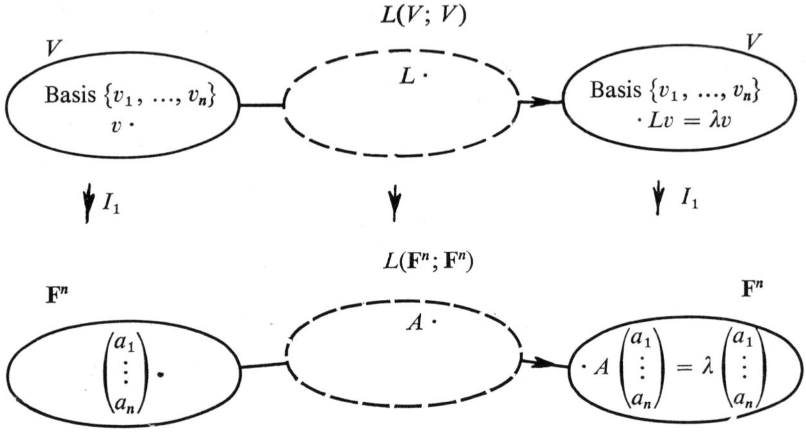


Proof: Let $v$ be an element of $V$ with representation $\sum_{i=1}^{n} a_i v_i$ relative to the given basis. Then $I_1$ is defined by $I_1 \colon v \mapsto (a_1, \ldots, a_n)$. If $v$ is an eigenvector of $L$ then $Lv = \lambda v$. Hence, since $L = I_1^{-1} A I_1$,

$$I_1^{-1} A I_1 v = \lambda v.$$

That is

$$A I_1 v = I_1 \lambda v,$$

giving

$$A(a_1, \ldots, {}_n a) = \lambda(a_1, \ldots, a_n).$$

Therefore $(a_1, \ldots, a_n)$ is an eigenvector of $A$ corresponding to the eigenvalue $\lambda$.

Similarly, using $A = I_1 L I_1^{-1}$, it can be seen that if $(a_1, \ldots, a_n)$ is an eigenvector of $A$ with eigenvalue $\lambda$, then $v$ is an eigenvector of $L$ with the same eigenvalue.

The problem of finding the eigenvectors and eigenvalues of an operator can thus be reduced to finding them for a matrix representation of the operator. This we have already discussed.

The student should now be able to see the significance of theorem 3, section 5.10, that eigenvalues are invariant under similarity transformations. If $A$ and $B$ are both matrix representations of $L$, then as we have seen in section 6.4, they are related by a similarity transformation and hence have the same eigenvalues. In this section we now see that each has the same eigenvalues as $L$.

If $A$ is the matrix representation of an operator $L$, then the *characteristic polynomial of* $L$ is defined to be that of $A$. This is a meaningful definition because, as can be seen by theorem 3, section 5.10, all matrix representations of $L$ have the same characteristic polynomial. The *characteristic equation of* $L$ is defined to be the characteristic equation of $A$.

*Example 1*

The set $\{1, x, x^2\}$ of functions on the interval $[0, 1]$ can be considered as a set of base vectors for a real three-dimensional real vector space $U$. Let $D = d/dx$ be an operator on $U$. Find the eigenvalues and eigenvectors of $D$.

We have already found the matrix representation of $D$ relative to this basis in example 1, section 6.1. We shall use the results and notation of that example.

The matrix representation of $D$ is $\begin{pmatrix} 0 & 1 & 0 \\ 0 & 0 & 2 \\ 0 & 0 & 0 \end{pmatrix}$. Hence the eigenvalues of $D$ are the eigenvalues of this matrix and the eigenvectors of $D$ and the matrix correspond under the isomorphism $I_1$.

There is one eigenvalue, namely 0, for the matrix. Hence $D$ has one eigenvalue, 0. The eigenvectors of the matrix corresponding to 0 are of the type $a(1, 0, 0)$, where $a$ is a non-zero real number. Under the isomorphism $I_1$ these correspond to the constant non-zero functions in $U$, namely functions of the type $f(x) = a1 + 0x + 0x^2$, $a \neq 0$. Hence the eigenvectors of $D$ corresponding to the eigenvalue 0 are the constant non-zero functions.

This result can be seen to be correct since $d/dx(f(x)) = 0f(x)$ if and only if $f(x)$ is a constant function.

The next theorem guarantees the existence of an eigenvector for an operator defined on a complex vector space.

*Theorem 2*

Let $L$ be an operator on a complex vector space $V$. Then there exists an eigenvector of $L$ in $V$.

Proof: Let $A$ be the matrix representation of $L$ relative to an arbitrary basis. The eigenvalues of $L$ and $A$ are thus identical and their eigenvectors correspond under the representation isomorphism. Theorem 2, section 5.10 ensures the existence of an eigenvector for $A$ in $C^n$. Hence there exists an eigenvector of $L$ in $V$.

Note that there is no guarantee of the existence of an eigenvector for an operator on a real vector space.

We now give a most important result that was mentioned earlier, namely a necessary and sufficient condition for a diagonal matrix representation to exist.

*Theorem 3*

An operator $L$ on a vector space $V$ can be represented by a diagonal matrix if and only if its eigenvectors span the space. The representation of $L$ relative to a basis consisting of eigenvectors is a diagonal matrix with the elements of the main diagonal being eigenvalues.

Proof: Let us assume that the eigenvectors of $L$ span $V$. Let $\{v_1, ..., v_n\}$ be a basis consisting of such eigenvectors. Then the matrix representation $A$ of $L$ relative to this basis is given by $(a_{ij})$, where

$$Lv_i = \sum_{j=1}^{n} a_{ji}v_j.$$

Let the eigenvalues corresponding to these base vectors be $\lambda_1, ..., \lambda_n$. Hence

$$Lv_i = \lambda_i v_i,$$

implying that

$$a_{ji} = \begin{cases} \lambda_i & \text{if } i = j, \\ o & \text{if } i \neq j. \end{cases}$$

Therefore the matrix representation of $L$ is

$$\begin{pmatrix} \lambda_1 & & O \\ & \ddots & \\ O & & \lambda_n \end{pmatrix}.$$

The student is asked to prove the converse in the exercises that follow.

The significance of the theorem telling us that a square $n \times n$ matrix $A$ is similar to a diagonal matrix $B$ if and only if it has $n$ linearly independent eigenvectors can now be seen (theorem 4, section 5.10). $A$ can be interpreted as an operator on $\mathbf{F}^n$. We have seen that $A$ is related to any other representation through a similarity transformation (section 6.4). The diagonal form $B$ is the matrix representation of the operator $A$ relative to a basis consisting of eigenvectors.

At this point we still do not know what kinds of operators have sufficient eigenvectors to span the space and thus have diagonal representation. In the next two sections we proceed to discuss special kinds of operators on unitary spaces. In further sections we shall find that normal operators do have diagonal matrix representations.

**Exercises**

1 Let $L$ be an operator acting on a vector space $V$ having an eigenvalue $\lambda$. Prove that the set consisting of all eigenvectors of $\lambda$ and the zero vector forms a subspace of $V$. This subspace is the eigenspace of $\lambda$ and is denoted $V_\lambda$.

2 If $L$ is an operator on a vector space $V$, prove that eigenspaces of $L$ corresponding to distinct eigenvalues intersect in the zero vector.

3 Prove that if an operator $L$ on a vector space $V$ has a diagonal matrix representation then its eigenvectors span $V$. Further prove that the diagonal representation is relative to a basis consisting of eigenvectors.

4 Let $L$ be an operator on a vector space $V$ that has representation by a diagonal matrix. Prove that the multiplicity of an eigenvalue as an element of the main diagonal of this representation is equal to its multiplicity as a root of the characteristic equation of $L$.

5 Let $V$ be the real vector space generated by the functions $e^x$ and $e^{-x}$ and $D^2 + 2D + 1$ be a differential operator on $V$. Determine the eigenvalues and eigenvectors of this operator. Prove that a diagonal matrix representation exists and determine it.

6 Determine the eigenvalues and eigenvectors of the projection operator $f$ of $\mathbf{R}^3$ into $\mathbf{R}^3$ defined by $f: (x, y, z) \mapsto (x, y, 0)$. Prove that $f$ has a diagonal matrix representation relative to the usual basis of $\mathbf{R}^3$ and interpret this result.

## 8.2 The Ajoint of an Operator. Self-Adjoint, Normal and Unitary Operators

In this section we shall assume that the vector space $V$ is a unitary space, that is, a vector space over the field of complex numbers or a subfield of this field, and that it has a Hermitian product which is positive-definite. Further we shall assume that the Hermitian product is non-degenerate.

Let $L$ be an operator on $V$. An operator $L^*$ on $V$ is said to be the *adjoint* of $L$ if, for all $u$ and $v$ in $V$,

$$\langle Lu, v\rangle = \langle u, L^*v\rangle .$$

*Theorem 1*

To each operator $L$ on a unitary space there exists a unique adjoint operator $L^*$.

Proof: Consider the inner product $\langle Lu, v\rangle$, where $v$ is a fixed but arbitrary vector and $u$ an arbitrary vector. The mapping $u \mapsto \langle Lu, v\rangle$ defines a functional on $V$ taking the vector $u$ into the scalar $\langle Lu, v\rangle$. Therefore there exists a vector $w$ such that $\langle Lu, v\rangle = \langle u, w\rangle$. The adjoint of $L$, $L^*$, is now defined by $L^*v = w$. It is easily seen that $L^*$ defined thus is an operator on $V$.

It remains to prove the uniqueness of $L^*$. Suppose there exist two such operators $L_1^*$ and $L_2^*$. Then

$$\langle Lu, v\rangle = \langle u, L_1^*v\rangle \quad \text{and} \quad \langle Lu, v\rangle = \langle u, L_2^*v\rangle ,$$

for arbitrary $u$ and $v$ in $V$. Hence

$$\langle u, L_1^*v\rangle = \langle u, L_2^*v\rangle ,$$

implying that

$$\langle u, (L_1^* - L_2^*)\, v\rangle = 0 .$$

By the non-degeneracy of the inner product, $(L_1^* - L_2^*)\, v = O$. Since $v$ is an arbitrary vector this implies that $L_1^* - L_2^*$ is the zero operator. Hence $L_1^* = L_2^*$.

*Theorem 2*

The mapping of an operator onto its adjoint $*: L \mapsto L^*$ is a bijective mapping of $L(V; V)$ onto itself that has the following properties:

a) $(L_1 + L_2)^* = L_1^* + L_2^*$;

b) $(aL)^* = \bar{a}L^*$;

c) $(L_1L_2)^* = L_2^*L_1^*$;

d) $L_1^{**} = L_1$.

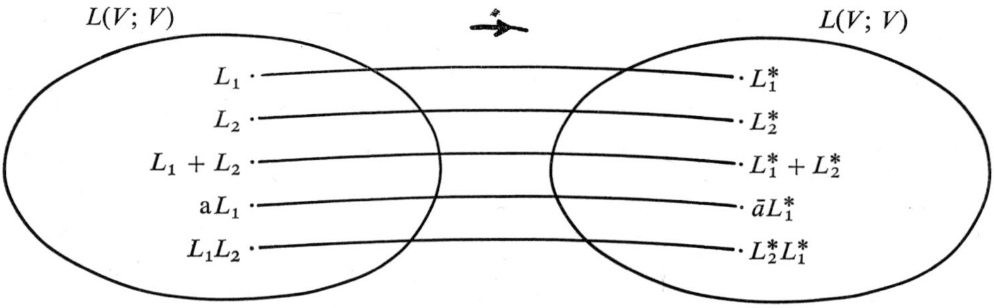


Proof: a) Let $u$ and $v$ be arbitrary elements of $V$.
Then

$$\begin{aligned}\langle (L_1 + L_2)\, u, v\rangle &= \langle L_1u + L_2u, v\rangle \\ &= \langle L_1u, v\rangle + \langle L_2u, v\rangle \\ &= \langle u, L_1^*v\rangle + \langle u, L_2^*v\rangle \\ &= \langle u, L_1^*v + L_2^*v\rangle \\ &= \langle u, (L_1^* + L_2^*)\, v\rangle .\end{aligned}$$

This implies that the adjoint of $(L_1 + L_2)$ is $(L_1^* + L_2^*)$. By the uniqueness of the adjoint of an operator we therefore have that $(L_1 + L_2)^* = L_1^* + L_2^*$.

The remaining sections of the theorem are proved using similar techniques and are left as exercises. Note that b) shows that the mapping $*$ is not in general linear. Only if the underlying field is the field of real numbers is the adjoint a linear mapping of $L(V; V)$ into $L(V; V)$.

*Example 1*

Let the matrix $A$ be interpreted as an operator on the space $\mathbf{R}^n$ with Hermitian product defined to be the dot product. Then the adjoint of $A$ is its transpose:

Let $X = (x_1, \ldots, x_n)$ and $Y = (y_1, \ldots, y_n)$ be arbitrary element of $\mathbf{R}^n$. Then the adjoint of $A$, $A^*$ is defined by

$$\langle AX, Y\rangle = \langle X, A^*Y\rangle .$$

This implies that $(AX) \cdot Y = X \cdot A^* Y$.

Here

$$(AX) . Y = \left[\sum_{j=1}^{n} a_{ij} x_j\right] . [y_k] = \sum_{i=1}^{n} \sum_{j=1}^{n} a_{ij} x_j y_i,$$

$$= [x_k] . \left[\sum_{i=1}^{n} a_{ij} y_i\right] = X . A^t Y.$$

Since the adjoint of $A$ is unique this implies that $A^* = A^t$.

*Example 2* p.83

Let the matrix $A$ be interpreted as an operator on the space $\mathbf{C}^n$ with Hermitian product defined to be the usual Hermitian product on this space. Then the adjoint of $A$ is the complex conjugate of its transpose:

Let $X = (x_1, \ldots, x_n)$ and $Y = (y_1, \ldots, y_n)$ be arbitrary elements of $\mathbf{C}^n$. Then the adjoint of $A$, $A^*$ is defined by

$$\langle AX, Y \rangle = \langle X, A^* Y \rangle.$$

This implies that $(AX) \cdot Y = X \cdot A^* Y$.

Here

$$(AX) \cdot Y = \left[\sum_{j=1}^{n} a_{ij} x_j\right] \cdot [y_k] = \sum_{i=1}^{n} \sum_{j=1}^{n} a_{ij} x_j \bar{y}_i$$

$$= [x_k] \cdot \left[\sum_{i=1}^{n} \bar{a}_{ij} y_i\right] = X \cdot (\overline{A^t})\, Y.$$

Since the adjoint of $A$ is unique this implies that $A^* = (\overline{A^t})$.

a matrix is Hermitian if it is equal to its transpose with all elements replaced by their complex conjugate p.24

Note that the adjoint of an operator is defined in terms of the Hermitian product of the space. The adjoint of a given operator will vary from Hermitian product to Hermitian product, being uniquely defined for each Hermitian product. The adjoints of the matrices take the above forms on $\mathbf{R}^n$ and $\mathbf{C}^n$ when these spaces have the Hermitian products given in the examples.

An operator $L$ on a unitary space $V$ is said to be *self-adjoint* if it is equal to its adjoint. That is, if

$$\langle Lu, v \rangle = \langle u, Lv \rangle$$

for all $u$ and $v$ in $V$.

*Theorem 3*

For a real vector space $V$ the self-adjoint operators in $L(V; V)$ form a real subspace of $L(V; V)$.

Proof: Let $L_1$ and $L_2$ be two self-adjoint operators. Then

$$\langle L_1 u, v \rangle = \langle u, L_1 v \rangle \quad \text{and} \quad \langle L_2 u, v \rangle = \langle u, L_2 v \rangle,$$

for every pair of elements $u$ and $v$ of $V$. Hence

$$\langle L_1u, v\rangle + \langle L_2u, v\rangle = \langle u, L_1v\rangle + \langle u, L_2v\rangle,$$

implying that

$$\langle (L_1 + L_2)\, u, v\rangle = \langle u, (L_1 + L_2)\, v\rangle,$$

proving that $L_1 + L_2$ is self-adjoint.

Further, for a real number $a$,

$$\langle aL_1u, v\rangle = a\langle L_1u, v\rangle = a\langle u, L_1v\rangle = \langle u, aL_1v\rangle,$$

proving that $aL_1$ in self-adjoint.

Note that we have limited the scalar field to the reals here. It can be seen that an analogous result does not hold for a complex vector space.

*Example 3*

Let the matrix $A$ be a self-adjoint operator on the space $\mathbf{R}^n$ with Hermitian product defined to be the dot product. Then $A$ is a symmetric matrix.

Since $A = A^*$ and $A^* = A^t$ it is seen that $A$ must be equal to its transpose and is therefore symmetric.

Conversely it can be seen that every symmetric matrix is a self-adjoint operator on this space.

*Example 4*

Let the matrix $A$ be a self-adjoint operator on the space $\mathbf{C}^n$ with Hermitian product defined to be the usual Hermitian product. Then the matrix $A$ is a Hermitian matrix.

Since $A = A^*$ and $A^* = (\overline{A^t})$ then $A = (\overline{A^t})$, proving the required result.

Conversely, it can be seen that every Hermitian matrix is a self-adjoint operator on this space.

An operator $N$ on a unitary space $V$ is said to be *normal* if it commutes with its adjoint, that is, if $NN^* = N^*N$. One significance of normal operators, as we shall see later in this chapter, is that they are operators on $V$ for which a diagonal matrix representation is possible.

*Theorem 4*

The set of all normal operators on $V$ contains all the self-adjoint operators and is closed under scalar multiplication.

Proof: It is obvious that every self-adjoint operator is normal. Further, if $N$ is normal, that is if $NN^* = N^*N$, then for a scalar $a$, since $(aN)^* = \bar{a}N^*$, $(aN)\,(aN)^* = (aN)\,(\bar{a}N^*) = a\bar{a}NN^* = \bar{a}aN^*N = (\bar{a}N^*)\,(aN) = (aN)^*\,(aN)$, proving that the set of all normal operators is closed under scalar multiplication.

It is left for the student to demonstrate that the set of normal operators is not closed under addition hence cannot form a subspace of $L(V; V)$.

Finally, let us discuss a special class of normal operators called *unitary operators*. An operator $U$ on a complex vector space is unitary if $UU^* = U^*U = I$, $I$ being the identity operator on the vector space. They are the normal operators with inverses being their adjoints. When the vector space is real, such operators are called *orthogonal operators*.

The significance of the unitary operators is that they are the operators that preserve all the structures of a unitary space; they are bijective and linear hence isomorphisms, preserving the linear structure; they preserve the norm and the Hermitian product. Mappings that preserve norms are called *isometries*.

We shall need the following lemma in the proof of this important result.

*Lemma* If $L$ is an operator on a unitary space $V$ such that $\langle Lv, v\rangle = 0$ for all $v \in V$, then $L = O$.

Proof: Let $u$ and $w$ be arbitrary elements of $V$ and $a$, $b$ arbitrary scalars. Then

$$\langle L(au + bw), au + bw\rangle - |a|^2 \langle Lu, u\rangle - |b|^2 \langle Lw, w\rangle = a\bar{b}\langle Lu, w\rangle + \bar{a}b\langle Lw, u\rangle.$$

By the given property of $L$ the left-hand side of this identity is zero, implying that

$$a\bar{b}\langle Lu, w\rangle + \bar{a}b\langle Lw, u\rangle = 0.$$

For $a = 1, b = 1$, this gives

$$\langle Lu, w\rangle + \langle Lw, u\rangle = 0,$$

and for $a = 1, b = i$,

$$-i\langle Lu, w\rangle + i\langle Lw, u\rangle = 0.$$

Solving this pair of simultaneous equations,

$$\langle Lu, w\rangle = 0$$

for all $u, w \in V$, implying that $L$ is the zero operator.

*Theorem 5*

An operator $U$ on a unitary space is unitary $\Leftrightarrow$ it is structure-preserving.

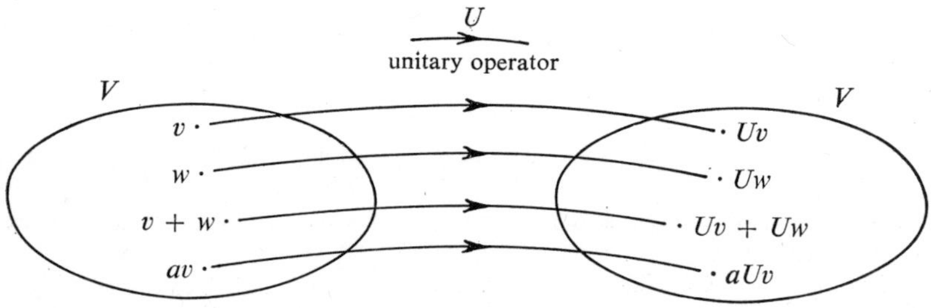

$$\|v\| = \|Uv\| \text{ preserves the norm}$$
$$\langle v, w\rangle = \langle Uv, Uw\rangle \text{ preserves the inner product.}$$

Proof: Let us first assume that $U$ is unitary, $UU^* = U^*U = I$.

If $v$ and $w$ are arbitrary elements of $V$, we have that

$$\|v\|^2 = \langle v, v\rangle = \langle U^*Uv, v\rangle = \langle Uv, Uv\rangle = \|Uv\|^2,$$

implying that

$$\|v\| = \|Uv\|;$$

$U$ preserves the norm.

Further,

$$\langle v, w\rangle = \langle U^*Uv, w\rangle = \langle Uv, Uw\rangle,$$

proving that $U$ preserves the inner product.

$U$, being an operator, is a linear mapping. We know that $U$ is bijective with inverse $U^*$ since it satisfies $UU^* = U^*U = I$ (exercise 10, section 3.1). Hence $U$ is an isomorphism. It thus preserves all the structure of the unitary space.

Conversely, if $U$ is structure-preserving, it is a bijective operator and hence its inverse exists and is an operator $U^{-1}$ satisfying $UU^{-1} = U^{-1}U = I$. Since it is an isometry

$$\|v\|^2 = \|Uv\|^2,$$

implying that

$$\langle v, v\rangle = \langle Uv, Uv\rangle,$$
$$\langle v, v\rangle = \langle U^*Uv, v\rangle,$$
$$\langle (U^*U - I)\, v, v\rangle = 0.$$

This result holds for arbitrary element $v$, hence by the previous lemma $U^*U - I = O$, implying that $U^*U = I$. We now have, since $U^{-1}$ exists, that $U^*UU^{-1} = IU^{-1}$, implying that $U^* = U^{-1}$. Hence $U$ is unitary.

## Exercises

1 Prove that if $L_1$ and $L_2$ are self-adjoint operators on a unitary space $V$ then their product $L_1L_2$ is self-adjoint $\Leftrightarrow L_1L_2 = L_2L_1$. (Note that this result tells us that the product of symmetric matrices is symmetric if and only if the matrices commute.) Hence the set of self-adjoint operators is not closed under the operation of composition.

2 Prove that if $N_1$ and $N_2$ are normal operators on $V$ with the property that either commutes with the adjoint of the other, then $N_1 + N_2$ and $N_1N_2$ are normal.

3 Consider the inner-product space of exercise 6 section 4.1. Find the adjoint of the matrix $\begin{pmatrix} 1 & 0 & 2 \\ 3 & 1 & 0 \\ 0 & 1 & 2 \end{pmatrix}$, considered as an operator on this space. Note that the adjoint is not the transpose, illustrating the remark at the end of example 2 of this section.

4 Let $\mathbf{C}^n$ be the complex vector space with Hermitian product defined by

$$\langle (x_1, \ldots, x_n), (y_1, \ldots, y_n) \rangle = x_1\bar{y}_1 + \ldots + x_n\bar{y}_n.$$

If $U$ is a unitary operator on $\mathbf{C}^n$, prove that the matrix representation of $U$ relative to the usual basis of $\mathbf{C}^n$ is a matrix whose columns are unit vectors which are mutually orthogonal. Prove that this result also holds for the rows of the matrix. Such a matrix is called a *unitary matrix*. The matrix representation of a unitary operator on $\mathbf{R}^n$, with Hermitian product defined to be the dot product, relative to the usual basis of $\mathbf{R}^n$, has similar properties. There the matrix is called an *orthogonal matrix*.

5 Extend the result of question 4 to show that if $V$ is a unitary space and $U$ a unitary operator of $V$, then any matrix representation of $U$ has the property that its columns or rows form an orthonormal set. Here the representation isomorphisms of $V$ onto $\mathbf{C}^n$ are also required to preserve the norm and Hermitian product to ensure that $V$ and any representation $\mathbf{C}^n$ have all structures identical.

6 Prove parts b), c) and d) of theorem 2 of this section.

## 8.3 Projection Operators

In section 3.2 an example was given of what we called a projection operator. This was an operator on $\mathbf{R}^3$ that projected the elements onto a subspace. In this section we define and develop the algebraic theory of such operators.

If $W$ is a subspace of a unitary space $V$, then

$$V = W \oplus W^{\perp}.$$

Hence if $v$ is an element of $V$ it can be expressed uniquely as the sum of an element $w$ of $W$ and an element $w'$ of $W^{\perp}$,

$$v = w + w'.$$

Since $w$ is uniquely determined by $v$, this decomposition of $v$ defines a mapping of $V$ into $W$,

$$Pv = w.$$

$P$ is seen to be a surjective operator onto $W$. It is called the *projection of $V$ onto $W$*. The vector $w$ is called the *projection of $v$ in $W$*. One can think of $w$ as being the component of $v$ in $W$. In this manner every subspace of $V$ determines a projection operator.

The orthogonal complement of $W$, $W^{\perp}$, being a subspace of $V$, will determine a projection operator $P'$,

$$P'(v) = w'.$$

It can be seen that the two projections $P$ and $P'$ are related:

$$P(v) + P'(v) = w + w',$$

$$(P + P')\,v = v,$$

$$\underline{P + P' = I,}$$

$I$ being the identity operator on $V$.

It should be noted that $W$ is the range of $P$, $W^{\perp}$ is the kernel of $P$ and that for $w \varepsilon W$, $P(w) = w$. We say that $P$ leaves the elements of its range invariant.

*Example 1*

Let us re-examine the example of section 3.2. $f$ is the operator of $\mathbf{R}^3$ into $\mathbf{R}^3$ defined by

$$f: (x, y, z) \mapsto (x, y, 0).$$

It is seen that

$$\mathbf{R}^3 = \underset{\substack{\uparrow \\ W}}{\{(x, y, 0) : x, y \,\varepsilon\, \mathbf{R}\}} \oplus \underset{\substack{\uparrow \\ W^{\perp}}}{\{(0, 0, z) : \; z \,\varepsilon\, \mathbf{R}\}}.$$

Hence $f$ satisfies our definition of a projection operator: it is the projection of $\mathbf{R}^3$ onto the two-dimensional subspace $\{(x, y, 0): x, y \,\varepsilon\, \mathbf{R}\}$. When $\mathbf{R}^3$ represents three-dimensional Euclidean space, $f$ is the projection of this space onto what is usually called the *xy-plane*.

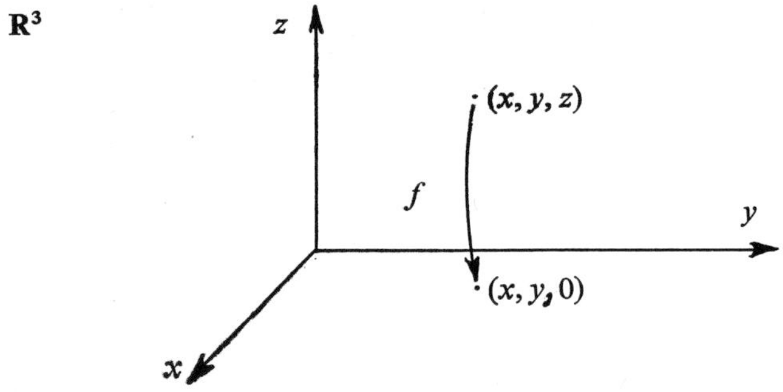


It can be seen that $f(x, y, 0) = (x, y, 0)$, illustrating the property of $f$ that it leaves elements of its range invariant. These elements already lie in the *xy plane*-the projection of any such element onto the *xy* plane is that element.

The following theorems give the properties of projection operators.

*Theorem 1*

A projection operator is *idempotent*. That is it has the property $P^2 = P$.
Proof: Let $P$ be the projection operator defined previously on $V$, with range $W$ and kernel $W^{\perp}$.

Then

$$Pv = w.$$

Furthermore,

$$P^2v = Pw = w,$$

giving

$$P^2v = Pv,$$

for arbitrary $v$ element of $V$, implying that $P$ is idempotent.

This property of the projection operator is to be expected, since on projection of the element once, the image lies in the required subspace. Further use of the projection operator will leave it unaltered.

*Theorem 2*

A projection operator is self-adjoint.
Proof: Let $P$ be the projection of $V$ onto $W$. If $v \varepsilon V$ then $v$ can be expressed uniquely

$$v = w + w'$$

where $w \varepsilon W$ and $w' \varepsilon W^{\perp}$. Hence,

$$\begin{aligned}\langle P^*v, v\rangle &= \langle v, Pv\rangle = \langle v, w\rangle = \langle w + w', w\rangle = \langle w, w\rangle + \langle w', w\rangle \\ &= \langle w, w\rangle = \langle w, w\rangle + \langle w, w'\rangle = \langle w, w + w'\rangle = \langle w, v\rangle = \langle Pv, v\rangle,\end{aligned}$$

giving

$$\langle P^*v, v\rangle = \langle Pv, v\rangle,$$

and hence

$$\langle (P^* - P)\, v, v\rangle = 0,$$

for all $v \varepsilon V$.

By the lemma in the previous section this implies that $P^* - P = O$, giving $P^* = P$.

The following theorem characterizes projection operators among all operators on a unitary space $V$.

*Theorem 3*

A necessary and sufficient condition for an operator $L$ on a unitary space $V$ to be a projection operator is that it be self-adjoint and idempotent.
Proof: We have already seen that a projection operator is both self-adjoint and idempotent. It remains to prove the converse.

Let $W$ be the range of $L$ and $U$ its kernel. The result will be proved if we can show that $U = W^{\perp}$, for then $L$ would be the projection operator of $V$ onto $W$. We shall first of all proceed to show that $V = W \oplus U$, using the idempotence of $L$.

An element of $W$ must be of the form $Lv$, for some $v \varepsilon V$. Let $Lv \varepsilon W \cap U$. The since $Lv \varepsilon U$, $L(Lv) = O$, implying that $L^2v = O$ and hence that $Lv = O$, by the idempotence of $L$. Therefore $W \cap U = \{O\}$.

Any element $v$ of $V$ can be expressed $v = Lv + (I - L)\, v$. $Lv$ is an element of $W$, and $L(I - L)\, v = (L - L^2)\, v = Ov = O$, implying that

$(I - L)\,v \,\varepsilon\, U$. Hence $V = W \oplus U$. This implies that the dimension of $U$ is equal to the dimension of $W^{\perp}$.

We now use the self-adjointness property of $L$ to show that $U \subseteq W^{\perp}$. Let $u \,\varepsilon\, U$ and $w \,\varepsilon\, W$. Then there exists an element $v$ of $V$ such that $Lv = w$. This gives,

$$\langle u, w\rangle = \langle u, Lv\rangle = \langle Lu, v\rangle = \langle O, v\rangle = 0.$$

Hence $U \subseteq W^{\perp}$.

Since the dimensions of these two spaces are identical $U = W^{\perp}$. Hence $L$ is a projection operator of $V$ onto its range $W$.

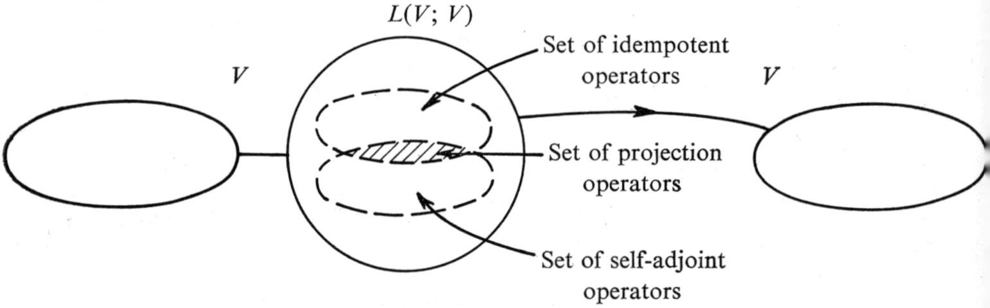


*A projection operator can now be redefined to be a self-adjoint, idempotent operator.*

*Theorem 4*

If $P_1$ and $P_2$ are projections onto subspaces $W_1$ and $W_2$ of $V$, then $W_1 \perp W_2 \Leftrightarrow P_1P_2 = O \Leftrightarrow P_2P_1 = O$.

Proof: Assume first of all that $W_1 \perp W_2$.

For $v$ an arbitrary element of $V$, $P_1P_2(v) = P_1w$, where $w \,\varepsilon\, W_2$. Since $W_1 \perp W_2$, $W_2 \subseteq W_1^{\perp}$. Hence $W_2$ is a subset of the kernel of $P_1$ and $P_1w = O$. This implies that $P_1P_2(v) = O$ for arbitrary $v$, element of $V$. Hence $P_1P_2 = O$.

Similarly $P_2P_1 = O$. We say that two projections $P_1$ and $P_2$ are *orthogonal* if and only if $P_1P_2 = O$.

Conversely if $P_1P_2 = O$ then for every $w$ element of $W_2$, $P_1w = P_1P_2w = O$, giving $W_2 \subseteq W_1^{\perp}$ and hence that $W_1 \perp W_2$.

*Theorem 5*

Let $P$ be a projection operator of $V$ onto $W$. Then if $v \,\varepsilon\, V$, $\|v\| \geqq \|Pv\|$, where the norm is the induced norm.

Proof: We have that

$$v = Pv + w',$$

where $Pv \varepsilon W$ and $w' \varepsilon W^{\perp}$. This gives

$$\|v\|^2 = \langle v, v\rangle = \langle Pv + w', Pv + w'\rangle = \langle Pv, Pv\rangle + \langle Pv, w'\rangle + \langle w', Pv\rangle + \langle w', w'\rangle.$$

But $\langle Pv, w'\rangle = \langle w, Pv\rangle = 0$ since $Pv \varepsilon W$ and $w' \varepsilon W^{\perp}$.
Therefore

$$\|v\|^2 = \|Pv\|^2 + \|w'\|^2,$$

giving

$$\|v\| \geqq \|Pv\|.$$

It can be seen that equality holds only when $w' = O$; that is when $v$ is originally in $W$, when of course $Pv = v$.

The theorem tells us that the norm of a vector is greater than or equal to the norm of its projection onto any subspace. This result is to be expected if we intuitively think of a projection of a vector onto a subspace as being its component in that subspace.

*Theorem 6*

If $P_1, \ldots, P_n$ are projections onto subspaces $W_1, \ldots, W_n$ of a unitary space $V$, then $P = P_1 + \cdots + P_n$ is a projection $\Leftrightarrow$ the $P_i$'s are pairwise orthogonal projections. If this is the case then $P$ is the projection of $V$ onto the subspace $W = W_1 \oplus \cdots \oplus W_n$. Further, if $W = V$ then $P_1 + \cdots + P_n = I$.

By pairwise orthogonal projections we mean that $P_iP_j = O$ if $i \neq j$.
Proof: Let us first assume that $P$ is a projection. For any $w \varepsilon W_i$ we have that $P_iw = w$. Using the idempotence and self-adjointness of projection operators,

$$\|w\|^2 = \|P_iw\|^2 = \langle P_iw, P_iw\rangle \leqq \sum_{k=1}^{n} \langle P_kw, P_kw\rangle = \sum_{k=1}^{n} \langle P_k^2w, w\rangle$$

$$= \sum_{k=1}^{n} \langle P_kw, w\rangle = \langle \sum_{k=1}^{n} P_kw, w\rangle = \langle Pw, w\rangle = \langle P^2w, w\rangle = \langle Pw, Pw\rangle$$

$$= \|Pw\|^2 \leqq \|w\|^2.$$

Hence equality must hold throughout, and

$$\sum_{k=1}^{n} \langle P_kw, P_kw\rangle = \langle P_iw, P_iw\rangle.$$

By the positive definiteness of the Hermitian product this implies that $P_jw = O$ for $j \neq i$.

Therefore for $v \varepsilon V$, $P_jP_iv = P_jw = O$, since $w \varepsilon W_i$, proving that $P_jP_i = O$ for $j \neq i$.

Conversely, if the projection operators are pairwise orthogonal, then for $v \varepsilon V$,

$$\begin{aligned} P^2 v &= (P_1 + \cdots + P_n)(P_1 + \cdots + P_n)\, v \\ &= (P_1^2 + \cdots + P_n^2)\, v \\ &= (P_1 + \cdots + P_n)\, v \\ &= Pv, \end{aligned}$$

giving that $P$ is idempotent.

Furthermore,

$$\begin{aligned} \langle Pv, v\rangle &= \langle (P_1 + \cdots + P_n)\, v, v\rangle = \sum_{k=1}^{n} \langle P_k v, v\rangle = \sum_{k=1}^{n} \langle v, P_k v\rangle \\ &= \langle v, \sum_{k=1}^{n} P_k v\rangle = \langle v, Pv\rangle, \end{aligned}$$

proving that $P$ is self-adjoint.

Hence $P$ is a projection.

Finally, to show that $P$ is the projection of $V$ onto the subspace $W = W_1 \oplus \cdots \oplus W_n$ we must show that $W$ is the range of $P$. We know that $w$ is an element of the range of $P$ if and only if $Pw = w$. Hence, if $w$ is an element of the range of $P$,

$$w = Pw = P_1 w + \cdots + P_n w.$$

Since $P_i w \varepsilon W_i$, $w \varepsilon W_1 \oplus \cdots \oplus W_n$, and the range of $P$ is a subset of $W$. Conversely if $w \varepsilon W_1 \oplus \cdots \oplus W_n$ then $w$ can be expressed in the form $w = w_1 + \cdots + w_n$, where $w_i \varepsilon W_i$. Thus

$$Pw = P(w_1 + \cdots + w_n) = w_1 + \cdots + w_n = w.$$

Hence $w$ is an element of the range of $P$, implying that $W$ is a subset of the range of $P$. It can therefore be concluded that the range of $P$ is $W$, proving that $P$ is a projection of $V$ onto $W$.

When $W = V$, $V = W_1 \oplus \cdots \oplus W_n$. Hence $v \varepsilon V$ can be expressed

$$v = w_1 + \cdots + w_n,$$

where $w_i \varepsilon V_i$.

This gives

$$v = P_1 v + \cdots + P_n v$$

implying that

$$Iv = (P_1 + \cdots + P_n)\, v,$$

$$I = P_1 + \cdots + P_n.$$

This is called a *resolution of the identity operator*.

*Example 2*

Let $\mathbf{R}^3$ be the unitary space with Hermitian product the usual dot product. Define three operators on $\mathbf{R}^3$ as follows:

$$P_1(x, y, z) = (x, 0, 0),$$

$$P_2(x, y, z) = (0, y, 0),$$

$$P_3(x, y, z) = (0, 0, z).$$

We shall show that each of these is a projection operator by using the second definition of a projection operator; that is that an operator is a projection operator if and only if it is self-adjoint and idempotent.

For $(x, y, z)$ and $(p, q, r)$ arbitrary elements of $\mathbf{R}^3$,

$$\langle P_1(x, y, z), (p, q, r)\rangle = \langle (x, 0, 0), (p, q, r)\rangle = xp = \langle (x, y, z), (p, 0, 0)\rangle$$
$$= \langle (x, y, z), P_1(p, q, r)\rangle.$$

Hence $P_1 = P_1^*$ and $P_1$ is self-adjoint. It can be proved in a similar manner that both $P_2$ and $P_3$ are self-adjoint.

Further, it can be seen that

$$P_1^2(x, y, z) = P_1(x, 0,0) = (x, 0, 0) = P_1(x, y, z),$$

implying that $P_1^2 = P_1$. Similarly $P_2$ and $P_3$ are idempotent. The three operators $P_1$, $P_2$ and $P_3$ are thus projection operators onto the subspaces $X = \{(x, 0, 0): x \varepsilon \mathbf{R}\}$, $Y = \{(0, y, 0): y \varepsilon \mathbf{R}\}$ and $Z = \{(0, 0, z): z \varepsilon \mathbf{R}\}$ of $\mathbf{R}^3$. These three projection operators are usually called the *coordinate projections* on $\mathbf{R}^3$ and the three subspaces are called the *coordinate axes.*

It can be seen that $X \perp Y \perp Z$ and it can easily be verified that $P_iP_j = O$ for $i \neq j$, illustrating theorem 4. Further, $P = P_1 + P_2 + P_3$ is the identity mapping of $\mathbf{R}^3$ onto $\mathbf{R}^3$. This is a projection operator of $\mathbf{R}^3$ onto $\mathbf{R}^3$, and $X \oplus Y \oplus Z = \mathbf{R}^3$, illustrating theorem 6.

By theorem 6, $P = P_1 + P_2$ is the projection of $\mathbf{R}^3$ onto the subspace $X \oplus Y$ of $\mathbf{R}^3$. In fact we have

$$P(x, y, z) = P_1(x, y, z) + P_2(x, y, z)$$
$$= (x, 0, 0) + (0, y, 0)$$
$$= (x, y, 0) \text{ which is an element of } X \oplus Y.$$

$P$ is thus the projection operator that was discussed in example 1. When $\mathbf{R}^3$ represents a three-dimensional Euclidean pace, $P_1$, $P_2$ and $P_3$ may be interpreted as projecting an arbitrary point $(x, y, z)$ onto the $x$, $y$ and $z$ axes.

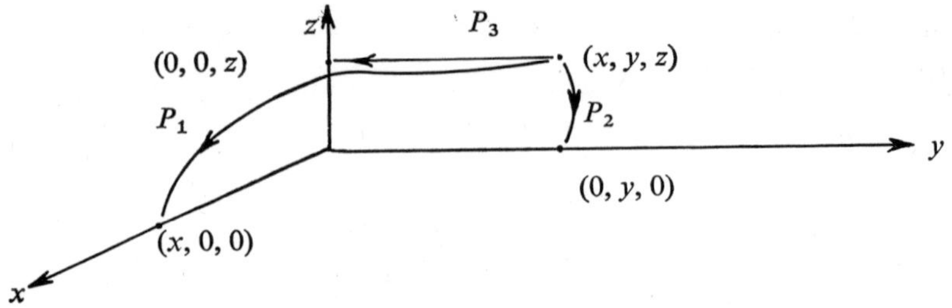


Here the subspaces $X$, $Y$ and $Z$ are the $x$, $y$ and $z$ axes respectively.

## Exercises

1 Prove that if $P$ is a projection operator of $V$ onto $W$ then $W$ is the range of $P$, $W^{\perp}$ is the kernel of $P$ and that for $w \varepsilon W$, $P(w) = w$.

2 Prove that the projection operators of example 2 of this section are pairwise orthogonal. Derive their matrix representations relative to the usual basis of $\mathbf{R}^3$. Note that these representations have all the properties of the operators hence can be interpreted as co-ordinate projections on $\mathbf{R}^3$.

3 If $P$ is a projection operator of $V$ onto $W$, a proper subspace, prove that it has eigenvalues $o$ and 1. Prove that the eigenvectors of $P$ span $V$ and hence that $P$ has a diagonal matrix representation. Note that a basis can always be chosen so that the diagonal matrix representation is $\begin{pmatrix} 1 & & & & O \\ & \ddots & & & \\ & & 1 & & \\ & & & 0 & \\ & & & & \ddots \\ O & & & & 0 \end{pmatrix}$. Every matrix of this type when interpreted as an operator is a projection operator.

4 Prove that the mapping $P$ of $\mathbf{R}^2$ into $\mathbf{R}^2$ defined by

$$P(x, y) = (x, 1)$$

is a projection operator. The Hermitian product on $\mathbf{R}^2$ is taken to be the usual dot product.

not linear ∴ not a projection operator

5 Let $V$ be the real inner product space spanned by $\sin t$ and $\cos t$, defined as real-valued functions on the interval $\left[-\frac{\pi}{2}, \frac{\pi}{2}\right]$ with Hermitian product defined by

$$\langle f, g\rangle = \int_{-\frac{\pi}{2}}^{\frac{\pi}{2}} fg\, dt.$$

Prove that the mapping $P$ of $V$ into $V$ defined by

$$P(a \sin t + b \cos t) = b \cos t$$

is a projection operator.

6 Define a projection $P_1$ of $\mathbf{R}^3$ that has as its range the subspace $A = \{a(1, 1, 0); a \varepsilon \mathbf{R}\}$ of $\mathbf{R}^3$. Similarly define a projection $P_2$ of $\mathbf{R}^3$ into $\mathbf{R}^3$ that has as its range the subspace $B = \{b(1, -1, 0): b \varepsilon \mathbf{R}\}$ of $\mathbf{R}^3$. Interpret these projections when $\mathbf{R}^3$ is taken to represent three-dimensional Euclidean space. Is $P = P_1 + P_2$ a projection? If so define its range.

7 Let $V$ be a unitary space with basis $\{u_1, \ldots, u_n\}$. Let $P_i$ be the projection operator onto the one-dimensional space $U_i$ having basis $u_i$. Prove that the projection of a vector $v$ onto this space is

$$\frac{\langle v, u_i\rangle}{\langle u_i, u_i\rangle} u_i.$$

It is then known from the theory of this section that

$$v = \frac{\langle v, u_i\rangle u_i}{\langle u_i, u_i\rangle} + w'$$

where $w' \in U_i^{\perp}$. Prove, using algebraic manipulation of inner products, that

$$\left\langle v - \frac{\langle v, u_i\rangle u_i}{\langle u_i, u_i\rangle}, au_i \right\rangle = o$$

for all scalars $a$, verifying this result. This is the general theory behind the construction of the orthogonal basis in the Gram-Schmidt orthogonalization process.

8 If $P$ and $Q$ are projection operators onto subspaces $U$ and $W$ of $V$, prove that $PQ$ is a projection $\Leftrightarrow PQ = QP$. Show that $PQ$ is a projection of $V$ onto $U \cap W$.

9 Prove that the intersection of the set of unitary operators and the set of projection operators on a unitary space $V$ is the set consisting of the single element, the identity operator.

## 8.4 Jordan Form of a Normal Operator

In this section we show that a basis exists for which a normal operator can be represented by a diagonal matrix. First we shall derive further properties of normal operators.

*Theorem 1*

Let $V$ be a unitary space and $N$ a normal operator on $V$. Then any eigenvector of $N$ with eigenvalue $\lambda$ is also an eigenvector of $N^*$ with eigenvalue $\bar{\lambda}$.

Proof: If $v$ is an arbitrary vector of $V$, then $\|N^*v\| = \|Nv\|$, for

$$\|N^*v\|^2 = \langle N^*v, N^*v\rangle = \langle v, NN^*v\rangle = \langle v, N^*Nv\rangle = \langle Nv, Nv\rangle = \|Nv\|^2.$$

If $N$ is normal then it can be seen that $N - \lambda I$ is normal, and

$$\|(N - \lambda I)\, v\| = \|(N - \lambda I)^*\, v\|,$$

implying that

$$\|(N - \lambda I)\, v\| = \|(N^* - \bar{\lambda} I)\, v\|.$$

If $v$ is an eigenvector of $N$ for the eigenvalue $\lambda$, then $(N - \lambda I)\, v = O$, implying that $\|(N - \lambda I)\, v\| = 0$ giving in turn that $\|(N^* - \bar{\lambda} I)\, v\| = 0$ and that $v$ is an eigenvector of $N^*$ for the eigenvalue $\bar{\lambda}$.

It is now convenient to introduce the concept of an invariant subspace. Let $V$ be a vector space, $L$ an operator on $V$ and $W$ a subspace of $V$. If for arbitrary $w \in W$, $Lw$ is also an element of $W$, $W$ is said to be an *invariant subspace of V under L*. We shall say that $W$ is an invariant subspace when the vector space and the operator are understood.

We have already come across examples of invariant subspaces. Any eigenspace of an operator $L$ is an invariant subspace and so also is the range. If $w$ is an element of an eigenspace $W$ of $L$ then it is an eigenvector and there exists a scalar $\lambda$ such that $Lw = \lambda w$. This implies that $L(Lw) = L(\lambda w) = \lambda(Lw)$, giving $Lw$ to be an eigenvector corresponding to the eigenvalue $\lambda$ and hence an element of $W$. For the range, if $v$ is an element of the range of $L$ then $Lv$ is also in the range.

It should be noted that if $W$ is invariant under an operator $L$ then the restriction of $L$ to $W$ is an operator on $W$. As such this restriction has an eigenvector in $W$. This vector will also be an eigenvector of $L$ in $W$.

*Theorem 2*

If $L$ is an operator on a finite-dimensional unitary space and $W$ is invariant under $L$, then $W^\perp$ is invariant under $L^*$.

Proof: Let $w \in W$ and $u \in W^\perp$. Then $Lw \in W$, and $\langle Lw, u\rangle = 0$. This implies that $\langle w, L^*u\rangle = 0$ giving that $L^*u$ is an element of $W^\perp$ whenever $u \in W^\perp$.

*Theorem 3*

If $N$ is a normal operator its eigenspaces are pairwise orthogonal. The projection operators onto these eigenspaces are pairwise orthogonal.

Proof: Let $v_i$ and $v_j$ be elements of distinct eigenspaces $W_{\lambda_i}$ and $W_{\lambda_j}$. Therefore $Lv_i = \lambda_i v_i$ and $Lv_j = \lambda_j v_j$.

Hence

$$\begin{aligned}\lambda_i\langle v_i, v_j\rangle &= \langle \lambda_i v_i, v_j\rangle\\ &= \langle Lv_i, v_j\rangle\\ &= \langle v_i, L^*v_j\rangle\\ &= \langle v_i, \bar{\lambda}_j v_j\rangle\\ &= \lambda_j\langle v_i, v_j\rangle.\end{aligned}$$

Since $\lambda_i \neq \lambda_j$ this implies that $\langle v_i, v_j\rangle = 0$, giving $W_{\lambda_i} \perp W_{\lambda_j}$. It is then known that the projection operators onto these eigenspaces are pairwise orthogonal (theorem 4, section 8.3).

*Theorem 4*

If $N$ is normal with eigenspaces $W_{\lambda_1}, \ldots, W_{\lambda_m}$, then its eigenvectors span $V$ and

$$V = W_{\lambda_1} \oplus \cdots \oplus W_{\lambda_m},$$

Proof: Let $W$ be the subspace of $V$ spanned by the eigenvectors of $N$. If $W$ is a proper subspace of $V$ then there exists a non-zero element $v$ of $V$ that is not an element of $W$. Since $V = W \oplus W^{\perp}$ then $v$ has a non-zero component in $W^{\perp}$ and $W^{\perp} \neq \{O\}$. But $W$ is an invariant subspace of $V$ under $N$, hence by theorem 2, $W^{\perp}$ is invariant under $N^*$. The restriction of $N^*$ to $W^{\perp}$ is thus an operator on $W^{\perp}$ and $N^*$ therefore has an eigenvector in $W^{\perp}$. But this would imply, by theorem 1, that $N$ has that same eigenvector in $W^{\perp}$ contradicting the assumption that all the eigenvectors of $N$ lie in $W$. Hence $W = V$.

We therefore have, since the eigenspaces are pairwise orthogonal and span $V$,

$$V = W_{\lambda_1} \oplus \cdots \oplus W_{\lambda_m}.$$

This decomposition of $V$ enables us to express a normal operator as a linear combination of the projection operators onto the eigenspaces:

If $v$ is an arbitrary element of $V$ it can be expressed uniquely in the form

$$v = w_1 + \cdots + w_m,$$

where $w_i \varepsilon W_{\lambda_i}$.

This gives

$$Nv = Nw_1 + \cdots + Nw_m$$

$$= \lambda_1 w_1 + \cdots + \lambda_m w_m.$$

Let $P_i$ be the projection of $V$ onto the subspace $W_i$. Then

$$Nv = \lambda_1 P_1 v + \cdots + \lambda_m P_m v$$

$$= (\lambda_1 P_1 + \cdots + \lambda_m P_m)\, v.$$

Since $v$ is an arbitrary element of $V$ this implies that

$$N = \lambda_1 P_1 + \cdots + \lambda_m P_m.$$

This resolution of the normal operator is called its *spectral resolution.* (This result together with the resolution of the identity operator [section 8.3] and the content of theorem 3 of this section make up what is sometimes called the *spectral theorem.*)

Further, it can be seen, on account of the pairwise orthogonality of the projection operators, that

$$N^2 = \lambda_1^2 P_1 + \cdots + \lambda_m^2 P_m,$$

and that for any positive integer $n$,

$$N^n = \lambda_1^n P_1 + \cdots + \lambda_m^n P_m.$$

If $p(N)$ is a polynomial in $N$ having complex coefficients, then by taking linear combinations of the above types of resolutions,

$$p(N) = p(\lambda_1)\,P_1 + \cdots + p(\lambda_m)\,P_m.$$

*Theorem 5*

If $N$ is normal then there exists an orthonormal basis for $V$ consisting of eigenvectors. The matrix representation of $N$ relative to this basis is a diagonal matrix with eigenvalues as elements of the main diagonal. This matrix representation of $N$ is called a *Jordan canonical form* for $N$.
Proof: Construct an orthonormal basis for each eigenspace using the Schmidt orthogonalization process. Since $V$ is the direct sum of the eigenspaces and the eigenspaces are pairwise orthogonal the set made up of bases is an orthonormal basis for $V$ made up of eigenvectors. Further, we know that the representation of $N$ relative to a basis of eigenvectors is a diagonal matrix with elements of the main diagonal being eigenvalues (Theorem 3, Section 8.1).

It should be mentioned in closing this section that since unitary operators and self-adjoint operators (which include projection operators), are all normal operators then they also have diagonal matrix representations and the other properties of normal operators discussed in this section.

## Exercises

1 Prove that the following matrices are normal when interpreted as operators on $\mathbf{R}^3$ with Hermitian product being the dot product. Find a Jordan canonical form in each case.

$$\text{a)} \begin{pmatrix} 3 & -1 & -1 \\ -2 & 2 & 0 \\ -3 & -1 & 3 \end{pmatrix} \qquad \text{b)} \begin{pmatrix} 2 & 0 & -1 \\ -1 & 1 & 0 \\ -2 & -1 & 2 \end{pmatrix}.$$

2 Prove that the converse of theorem 3 holds.

3 Prove that the converse of theorem 4 holds.

4 Prove that the converse of theorem 5 holds.

5 Prove that an operator on a unitary space, having a diagonal matrix representation is normal.

6 Let $V$ be a unitary space and $P$ a projection operator of $V$ onto a subspace $W$. Since $P$ is self-adjoint it is normal. Show that a Jordan form of $P$ exists of the type

$$\begin{pmatrix} 1 & & & & & & O \\ & 1 & & & & & \\ & & \ddots & & & & \\ & & & 1 & & & \\ & & & & 0 & & \\ & & & & & \ddots & \\ O & & & & & & 0 \end{pmatrix}.$$

Show that here the number of ones in the diagonal will correspond to the dimension of the range of $P$.

7 Let $N$ be a normal operator which has a Jordan form

$$\begin{pmatrix} 1 & 0 & 0 & 0 \\ 0 & 1 & 0 & 0 \\ 0 & 0 & 2 & 0 \\ 0 & 0 & 0 & 3 \end{pmatrix}.$$

Using the theorems of this section write down the eigenvalues of $N$ and determine by inspection the dimension of each eigenspace.

8 Prove that the multiplicty of the eigenvalue $\lambda_i$ in the Jordan form of a normal operator $N$ is equal to its multiplicity as a root of the characteristic equation of $N$.

9 If $N$ is a normal operator and $P_1, \ldots, P_m$ are the projection operators on to the eigenspaces, prove that a polynomial $p(N)$ in $N$, having complex coefficients, may be written

$$p(N) = p(\lambda_1) P_1 + \cdots + p(\lambda_m) P_m.$$

## 8.5 Generalized Eigenvectors

In the previous section we defined and discussed the properties of eigenvectors of normal operators. We found that since there are sufficient eigenvectors to form a basis for the space they can be used to lead to a diagonal matrix representation of the operator. When an operator is such that there are insufficient eigenvectors to span the space we shall see that it is possible in the case of complex vector spaces to complete a basis with generalized eigenvectors and obtain a particularly simple representation called its Jordan canonical form. It is the purpose of this section and the next to define and analyze some of the properties of generalized eigenvectors.

Let $V$ be a vector space of dimension $n$ over a field $\mathbf{F}$ and let $L$ be an operator on $V$. The set of vectors in $V$ for which there exists a positive integer $k$ such that

$$L^k(v) = O,$$

is called the *generalized kernel* or *generalized null space* of the operator. It can be seen that this is in fact a generalization of the concept of the kernel. An eigenvector of $L$ is a non-zero vector $v$ for which there exists a scalar $\lambda$ such that

$$(L - \lambda I)v = O.$$

Here $I$ is the identity operator on the space. The set of eigenvectors corresponding to the eigenvalue $\lambda$ is thus the non-zero subset of the kernel of $(L - \lambda I)$. A natural generalization of the eigenvector is the generalized eigenvector defined thus:

A non-zero vector $v$ is a *generalized eigenvector* corresponding to the eigenvalue $\lambda$ if there exists a positive integer $k$ such that

$$(L - \lambda I)^k v = O.$$

That is, a non-zero vector $v$ is an eigenvector of $L$ if there exists a scalar $\lambda$ such that $v$ is an element of the kernel of $(L - \lambda I)$ and $v$ is a generalized eigenvector if it is a non-zero element of the generalized kernel of $(L - \lambda I)$.

A generalized eigenvector $v$ is said to be of *rank* $k$ if

$$(L - \lambda I)^k v = O$$

and

$$(L - \lambda I)^{k-1} v \neq O.$$

Thus eigenvectors are generalized eigenvectors of rank 1.

*Theorem 1*

Let $V$ be a vector space and $L$ an operator on $V$. If $\lambda$ is an eigenvalue of $L$ then the set consisting of the generalized eigenvectors corresponding to $\lambda$, and the zero vector, forms a subspace of $V$ called the *generalized eigenspace* for the eigenvalue $\lambda$.

Proof: Let $N$ be the set consisting of generalized eigenvectors corresponding to the eigenvalue $\lambda$, and the zero vector. For $v_1$ and $v_2$ non-zero elements of $N$ there exist positive integers $k_1$ and $k_2$ such that

$$(L - \lambda I)^{k_1} v_1 = O \quad \text{and} \quad (L - \lambda I)^{k_1 - 1} v_1 \neq O,$$

$$(L - \lambda I)^{k_2} v_2 = O \quad \text{and} \quad (L - \lambda I)^{k_2 - 1} v_2 \neq O.$$

Hence if $k$ is the larger of $k_1$ and $k_2$,

$$(L - \lambda I)^k (v_1 + v_2) = O,$$

proving that $(v_1 + v_2)$ is an element of $N$. Further, for an arbitrary scalar $a$, $av_1$ is an element of $N$. Hence $N$ is a subspace.

It can be seen that the eigenspace corresponding to the eigenvalue $\lambda$ is a subspace of the generalized eigenspace of $\lambda$. This may or may not be a proper subspace as the following example illustrates.

*Example 1*

Find the generalized eigenvectors of the matrix $A = \begin{pmatrix} 1 & 3 & 3 \\ 0 & 1 & 2 \\ 0 & 0 & 4 \end{pmatrix}$ when $A$ is interpreted as a linear operator on $\mathbf{R}^3$.

One finds that the eigenvalues of $A$ are 1 (repeated) and 4. Corresponding to the eigenvalue 1 there is a one-dimensional eigenspace with basis $\{(1, 0, 0)\}$ and corresponding to the eigenvalue 4 there is a one-dimensional eigenspace with basis $\{(5, 2, 3)\}$. Hence one can only construct two linearly independent eigenvectors of $A$ in $\mathbf{R}^3$. We now search for generalized eigenvectors corresponding to the eigenvalue 1. A vector $v$ will be a generalized eigenvector corresponding to the eigenvalue 1 if there exists a positive integer $k$ such that $(A - I_3)^k v = O$.

For $k = 2$, $(A - I_3)^2 v = O \Rightarrow \begin{pmatrix} 0 & 0 & 18 \\ 0 & 0 & 8 \\ 0 & 0 & 16 \end{pmatrix} \begin{pmatrix} a \\ b \\ c \end{pmatrix} = O$.

Any vector of the type $(a, b, 0)$, where $a$ and $b$ are real numbers satisfies this condition. On investigating the condition $(A - I_3)^k v = O$ with further positive integers $k$ one obtains no new vectors. For example, for $k = 3$,

$$(A - I_3)^3 v = O \Rightarrow \begin{pmatrix} 0 & 0 & 102 \\ 0 & 0 & 32 \\ 0 & 0 & 64 \end{pmatrix} \begin{pmatrix} p \\ q \\ r \end{pmatrix} = O \Rightarrow \begin{pmatrix} 0 & 0 & 18 \\ 0 & 0 & 8 \\ 0 & 0 & 16 \end{pmatrix} \begin{pmatrix} p \\ q \\ r \end{pmatrix} = O.$$

Hence the generalized eigenspace corresponding to the eigenvalue 1 is the set $\{v: v = (a, b, 0): a, b \in \mathbf{R}\}$. It can be seen that the eigenspace corresponding to the eigenvalue is a proper subspace of this space. The elements of the eigenspace are generalized eigenvectors of rank 1. Any other element of the generalized eigenspace is a generalized eigenvector of rank 2. For example the vector $(1, 1, 0)$ is a generalized eigenvector of rank 2 since it satisfies the conditions $(A - I_3)^2 v = O$, $(A - I_3) v \neq O$. There do not exist generalized eigenvectors of rank other than one or two corresponding to the eigenvalue 1.

For the eigenvalue 4 the generalized eigenvectors are the vectors that satisfy $(A - 4I_3)^k v = O$ for some positive integer $k$. The eigenspace, that is the set of vectors that satisfy the above condition with $k = 1$, is one-dimensional, spanned by the vector $(5, 2, 3)$. Further values of $k$ lead to no new vectors. Hence the eigenvectors are the only generalized eigenvectors for the eigenvalue 4. Here the generalized eigenspace reduces to the eigenspace.

## Exercises

1 Find the generalized eigenvectors and generalized eigenspaces of the following matrices:

$$\text{a)} \begin{pmatrix} 1 & 1 & 0 \\ 0 & 1 & 0 \\ 0 & 0 & 1 \end{pmatrix} \qquad \text{b)} \begin{pmatrix} 2 & 1 & 0 \\ 0 & 2 & 1 \\ 0 & 0 & 2 \end{pmatrix}.$$

Here the matrices are interpreted as linear mappings of $\mathbf{R}^3$ into $\mathbf{R}^3$.

2 Find the generalized eigenvectors and the generalized eigenspaces of the matrices

$$\text{a)} \begin{pmatrix} 2 & 1 & 0 & 0 \\ 0 & 2 & 0 & 0 \\ 0 & 0 & 3 & 1 \\ 0 & 0 & 0 & 3 \end{pmatrix} \qquad \text{b)} \begin{pmatrix} 4 & 1 & 0 & 0 \\ 0 & 4 & 0 & 0 \\ 0 & 0 & 2 & 0 \\ 0 & 0 & 0 & 2 \end{pmatrix} \qquad \text{c)} \begin{pmatrix} 2 & 0 & 4 & 3 \\ 0 & 2 & 0 & 2 \\ 0 & 0 & 1 & 1 \\ 0 & 0 & 0 & 1 \end{pmatrix},$$

when the matrices are interpreted as linear mappings of $\mathbf{R}^4$ into $\mathbf{R}^4$.

## 8.6 Further Properties of Generalized Eigenvectors

In this section we analyze further the properties of generalized eigenvectors and generalized eigenspaces. The properties of the generalized eigenspaces will enable us to select generalized eigenvectors that will later be used as base vectors to give a canonical matrix representation of the linear operator.

*Theorem 1*

Let $L$ be an operator on a vector space $V$ and $\lambda$ an eigenvalue of $L$. Generalized eigenvectors of $L$ corresponding to $\lambda$ but of differing ranks are linearly independent.

Proof: Let $v_p, \ldots, v_q$ be generalized eigenvectors of differing ranks $p, \ldots, q$ respectively, all corresponding to $\lambda$ and arranged with $p < \cdots < q$.

Consider the identity

$$a_p v_p + \cdots + a_q v_q = O,$$

for arbitrary scalars $a_p, \ldots, a_q$.

This identity implies that

$$(L - \lambda I)^{q-1} (a_p v_p + \cdots + a_q v_q) = O.$$

By the linearity of $(L - \lambda I)^{q-1}$, (it is the composite of $q - 1$ linear mappings), this gives

$$a_q(L - \lambda I)^{q-1} v_q = O,$$

implying that $a_q$ is identically zero. Further it can be seen that all the remaining coefficients in the original identity are similarly, in turn, found to be zero, giving the generalized eigenvectors to be linearly independent.

*Theorem 2*

Let $L$ be an operator on a vector space $V$, $\lambda$ an eigenvalue of $L$ and $N$ the generalized eigenspace corresponding to $\lambda$. If $v$ is a generalized eigenvector of rank $m, m > 1$, in $N$, then $(L - \lambda I) v$ is a generalized eigenvector of rank $m - 1$ in $N$.

Proof: Since $v$ is a generalized eigenvector of rank $m$,

$$(L - \lambda I)^m v = O,$$

and

$$(L - \lambda I)^{m-1} v \neq O.$$

This implies that

$$(L - \lambda I)^{m-1} (L - \lambda I) v = O,$$

and

$$(L - \lambda I)^{m-2} (L - \lambda I) v \neq O,$$

proving that $(L - \lambda I) v$ is a generalized eigenvector of rank $m - 1$ in $N$.

Further it can be seen that $(L - \lambda I)^2 v$ is a generalized eigenvector of rank $m - 2$ in $N$, for

$$(L - \lambda I)^{m-2} (L - \lambda I)^2 v = O,$$

and

$$(L - \lambda I)^{m-3} (L - \lambda I)^2 v \neq O.$$

Hence a generalized eigenvector $v$ of rank $m$, $m > 1$, in $N$ can in this manner be used to generate a *chain* of $m$ linearly independent generalized eigenvectors of ranks $m, m - 1, \ldots, 1$ in $N$. They are the vectors

$$v, (L - \lambda I) v, (L - \lambda I)^2 v, \ldots (L - \lambda I)^{m-1} v.$$

The final vector in the chain, namely $(L - \lambda I)^{m-1} v$ will be an eigenvector since it is a non-zero vector that satisfies the identity

$$(L - \lambda I)(L - \lambda I)^{m-1} v = O.$$

A chain is said to be a *maximal chain* if its elements cannot be considered a proper subset of another chain. Maximal chains must exist since the vectors that make up a chain are linearly independent and the generalized eigenspace in which they lie is of finite dimension.

Let $r$ be the largest rank of any generalized eigenvector in $N$. Then the *index* of $N$ is said to be $r$. Note that this implies that there exists a maximal chain in $N$ consisting of $r$ vectors. There may be more than one maximal chain in $N$ consisting of $r$ vectors and there may be maximal chains in $N$ consisting of fewer than $r$ vectors. It can be seen that if $r$ is the index of $N$ then $N$ is the kernel of $(L - \lambda I)^r$. That is, $(L - \lambda I)^r N = O$. (Every element of this kernel is a generalized eigenvector corresponding to $\lambda$ and every generalized eigenvector is an element of this kernel.)

*Theorem 3*

Let $V$ be a vector space of dimension $n$, $L$ an operator on $V$ and $\lambda$ an eigenvalue of $L$. If $N$ is the generalized eigenspace corresponding to $\lambda$, $r$ is the index of $N$ and $W$ the range of the operator $(L - \lambda I)^r$, then $V = N \oplus W$.
Proof: $W$ is the *generalized range* of $L$ for the eigenvalue $\lambda$. We shall first of all prove that $N \cap W = \{O\}$. Let $v \in N \cap W$. Then $v \in N \Rightarrow (L - \lambda I)^r v = O$, and $v \in W \Rightarrow v = (L - \lambda I)^r u$ for some $u \in V$. Thus, combining these two statements,

$$(L - \lambda I)^r (L - \lambda I)^r u = O,$$

implying that $u$ is a generalized eigenvector. But since $r$ is the largest rank of any generalized eigenvector corresponding to $\lambda$, this implies

$$(L - \lambda I)^r u = O.$$

Hence $v = O$, proving that $N \cap W = \{O\}$.

Further, let $\{n_1, \ldots, n_p\}$ and $\{w_1, \ldots, w_q\}$ be bases for $N$ and $W$ respectively. If $n_i$ is one of the base vectors for $N$ then it is linearly independent of $w_1, \ldots, w_q$, for otherwise it would be a non-zero vector in both $N$ and $W$ violating $N \cap W = \{O\}$. Hence $\{n_1, \ldots, n_p, w_1, \ldots, w_q\}$ is a set of $p + q$ linearly independent vectors in $V$. But by theorem 7, section 3.4, $\dim V = \dim N + \dim W$. Hence $n = p + q$. Thus the set $\{n_1, \ldots, n_p, w_1, \ldots, w_q\}$ is a set of $n$ linearly independent vectors in $V$ and therefore forms a basis for $V$. This implies that $V = N + W$ which together with the condition $N \cap W = \{O\}$ proves the theorem.

*Theorem 4*

The generalized eigenspace $N$ and generalized range $W$ of an operator $L$, corresponding to an eigenvalue $\lambda$, are both invariant subspaces.

Proof: If $v \varepsilon N$ then $(L - \lambda I) v \varepsilon N$ by theorem 2. This proves that $N$ is an invariant subspace under the operator $(L - \lambda I)$. Further if $(L - \lambda I) v \varepsilon N$ then $Lv - \lambda v \varepsilon N$, implying that $Lv \varepsilon N$ since $N$ is a vector space. Hence $N$ is invariant under $L$. It is left as an exercise to prove that the generalized range of $L$ is an invariant subspace.

It should be remembered that if $U$ is an invariant subspace of $V$ under an operator $L$ then the restriction of $L$ to $U$ is an operator on $U$. As such the restriction of $L$ will have a matrix representation relative to a basis in $U$. Two exercises are given which illustrate how such a matrix representation is embedded in the matrix representation of $L$.

## Exercises

1 Give the index of each generalized eigenspace for the matrices in exercises 1 and 2 of section 8.5. Give an example of a maximal chain in each generalized eigenspace.

2 If $U$ is a proper invariant subspace of $V$ under an operator $L$ prove that a matrix representation of $L$ exists of the form

$$\begin{pmatrix} A & B \\ O & C \end{pmatrix}.$$

3 If $V$ is the direct sum of two subspaces $U$ and $W$ prove that a matrix representation of the operator exists, of the form

$$\begin{pmatrix} A & O \\ O & B \end{pmatrix}$$

where $A$ and $B$ are matrix representations of the restrictions of the operator to the subspaces.

4 Prove that the generalized range $W$ of an operator $L$, corresponding to an eigenvalue $\lambda$, is an invariant subspace.

## 8.7 Jordan Blocks

Here we demonstrate how a suitable basis for the vector space can be selected to give a matrix representation of an operator in what we call a Jordan block form. It will be assumed that the vector space is over the field of complex numbers. We have existence theorems for eigenvectors of operators on such vector spaces, which we shall use. The theory does not hold for real vector spaces.

If $V$ is a complex vector space, $L$ an operator on $V$ and $N_1$, $W_1$ the generalized eigenspace and generalized range for an eigenvalue $\lambda_1$ then we have seen that $V = N_1 \oplus W_1$ and that $N_1$ and $W_1$ are both invariant subspaces. As previously, let $\{n_1, \ldots, n_p\}$ and $\{w_1, \ldots, w_q\}$ be bases for $N_1$ and $W_1$ respectively, giving $\{n_1, \ldots, n_p, w_1, \ldots, w_q\}$ to be a basis for $V$. Relabel these vectors $\{v_1, \ldots, v_p, v_{p+1}, \ldots, v_n\}$ so that $v_1, \ldots, v_p$ in the basis for $N_1$ and $v_{p+1}, \ldots, v_n$ is the basis for $W_1$.

To find the matrix representation of $L$ relative to this basis we examine the effect of $L$ on the base vectors. Let

$$Lv_i = \sum_{j=1}^{n} a_{ji} v_j.$$

Since $N_1$ is invariant under $L$, $Lv_i$ is an element of $N_1$ for $1 \leqq i \leqq p$. Hence $a_{ji} = 0$ for $1 \leqq i \leqq p$ and $j > p$. Similarly since $W_1$ is invariant under $L$, $Lv_i$ is an element of $W_1$ for $p + 1 \leqq i \leqq n$, giving $a_{ji} = 0$ for $p + 1 \leqq i \leqq n$ and $j < p + 1$. The matrix of coefficients $(a_{ji})$ is thus of the following form

$$\begin{pmatrix} a_{11} & \ldots & a_{1p} & 0 & \ldots & 0 \\ \vdots & & \vdots & \vdots & & \vdots \\ a_{p1} & \ldots & a_{pp} & 0 & \ldots & 0 \\ 0 & \ldots & 0 & a_{(p+1)(p+1)} & & a_{(p+1)n} \\ \vdots & & \vdots & \vdots & & \vdots \\ 0 & \ldots & 0 & a_{n(p+1)} & & a_{nn} \end{pmatrix}.$$

The transpose of this matrix will be the matrix representation of $L$ relative to this basis. It can be seen that it will be a matrix with the same type of structure, namely

$$\begin{pmatrix} A_1 & O \\ O & B_1 \end{pmatrix},$$

where $A_1$ is a $p \times p$ matrix and $B_1$ is a $q \times q$ matrix. $A_1$ is the matrix representation of the restriction of $L$ to $N_1$, relative to the basis $\{v_1, \ldots, v_p\}$ of $N_1$. $B_1$ is the matrix representation of the restriction of $L$ to $W_1$, relative to the basis $\{v_{p+1}, \ldots, v_n\}$. The above structure of the matrix representation is independent of the choice of bases for $N_1$ and $W_1$.

Consider the restriction of $L$ to $W_1$. Since $W_1$ is invariant under $L$ this restriction is an operator on $W_1$. Hence by theorem 2, section 8.1, there exists an eigenvalue for this mapping. This eigenvalue, $\lambda_2$, must be distinct from the previous eigenvalue, for otherwise an eigenvector that corresponds to $\lambda_2$ would be in both $N_1$ and $W_1$ contradicting $N_1 \cap W_1 = \{O\}$. Let $N_2$ be the generalized eigenspace of $\lambda_2$ and $W_2$ the generalized range. Then $N_2$ and $W_2$ are invariant subspaces of $W_1$ under the restriction of $L$ to $W_1$, and are such that $W_1 = N_2 \oplus W_2$. $N_2$ and $W_2$ will also be invariant subspaces of $V$ under $L$ with

$$V = N_1 \oplus N_2 + W_2.$$

The process is now continued, $W_2$ must contain an eigenvector of $L$. Since $V$ is of finite dimension this procedure will terminate when $W_i = \phi$. Hence $V$ can be represented

$$V = N_1 \oplus N_2 + \cdots \oplus N_h,$$

where $N_i$ is the generalized eigenspace for the eigenvalue $\lambda_i$. A finite-dimensional complex vector space can thus always be spanned by generalized eigenvectors.

On choosing bases for these spaces and combining the sets of bases while preserving order, to give a basis for $V$, it can be seen that the matrix representation of $L$ relative to this basis is

$$A = \begin{pmatrix} A_1 & & & & \\ & A_2 & & O & \\ & & \ddots & & \\ & O & & A_i & \\ & & & & \ddots \\ & & & & & A_h \end{pmatrix}.$$

Here $A_i$ is the matrix representation of the restriction of $L$ to the generalized eigenspace $N_i$. The matrices $\{A_i\}$ are called *Jordan blocks* of $A$.

This overall structure of $A$ is independent of the choice of base vectors in each generalized eigenspace. By a special selection of the base vectors in the generalized null space $N_i$ we demonstrate in the next section that $A_i$ can be put into a particularly simple form. We do this for each generalized null space to get the canonical matrix representation of $L$.

*Example*

Find a Jordan block form of the matrix $A = \begin{pmatrix} 1 & 3 & 3 \\ 0 & 1 & 2 \\ 0 & 0 & 4 \end{pmatrix}$, when the matrix is interpreted as a linear operator on $\mathbf{R}^3$.

This operator was discussed in section 8.5. It was found that there existed two eigenvalues 1 and 4 with generalized eigenspaces

$$N_1 = \{(a, b, 0) \colon a, b \in \mathbf{R}\} \quad \text{and} \quad N_2 = \{c(5, 2, 3) \colon c \in \mathbf{R}\}$$

respectively.

It can be seen that $\mathbf{R}^3 = N_1 \oplus N_2$. To obtain a Jordan block form of the operator we select a basis for $\mathbf{R}^3$ consisting of base vectors for $N_1$ followed by base vectors for $N_2$. Note that according to the theory it is essential to keep the sets together in ordering the basis for $\mathbf{R}^3$. Hence the basis $\{(1, 0, 0), (0, 1, 0), (5, 2, 3)\}$ should lead to a Jordan block form of the operator.

We see that relative to this basis the matrix representation is

$$\left(\begin{array}{cc|c} 1 & 1 & 0 \\ 0 & 1 & 0 \\ \hline 0 & 0 & 4 \end{array}\right),$$

a Jordan block form. Note that the matrix $\begin{pmatrix} 1 & 1 \\ 0 & 1 \end{pmatrix}$ is the matrix representation of the restriction of $A$ to the subspace $N_1$, relative to the basis $\{(1, 0, 0), (0, 1, 0)\}$ of $N_1$. (4) is the matrix representation of the restriction of $A$ to the subspace $N_2$, relative to the basis $\{(5, 2, 3)\}$ of $N_2$.

The theory developed here is used in determining solutions to systems of linear differential equations having constant coefficients. If $X' = AX$ is such a system (section 2.2), $A$ is a matrix with generalized eigenspaces $N_1, \ldots, N_k$ of dimensions $n_1, \ldots, n_k$ respectively then it can be shown that the solution $X(t)$ satisfying a given initial $X(0)$ is

$$X(t) = \sum_{j=1}^{k} e^{\lambda_j t} \sum_{i=0}^{n_j - 1} \frac{t^i}{i!} (A - \lambda_j I)^i v_j, \quad (-\infty < t < \infty),$$

where $v_1, \ldots, v_k$ are unique vectors in $N_1, \ldots, N_k$ such that $X(0) = v_1 + \cdots + v_k$.[1]

## Exercises

1 Find Jordan block forms for the following operators on $\mathbf{R}^3$:

a) $\begin{pmatrix} 2 & 1 & 4 \\ 0 & 2 & 3 \\ 0 & 0 & 2 \end{pmatrix}$; b) $\begin{pmatrix} 1 & 2 & 3 \\ 0 & 4 & 1 \\ 0 & 0 & 4 \end{pmatrix}$.

[1] A discussion of this method is given in F. Brauer and J. A. Nohel, Qualitative Theory of Ordinary Differential Equations, section 2–5 (W. A. Benjamin Inc., 1969).

2 Find Jordan block forms for the following operators:

a) $\begin{pmatrix} 2 & 0 & 4 & 3 \\ 0 & 2 & 0 & 2 \\ 0 & 0 & 1 & 1 \\ 0 & 0 & 0 & 1 \end{pmatrix}$, on $\mathbf{R}^4$; b) $\begin{pmatrix} 4 & 3 & 2 & 4 & 1 \\ 0 & 4 & 1 & 0 & 1 \\ 0 & 0 & 4 & 2 & 2 \\ 0 & 0 & 0 & 1 & 3 \\ 0 & 0 & 0 & 0 & 1 \end{pmatrix}$, on $\mathbf{R}^5$;

c) $\begin{pmatrix} 3 & 1 & 2 & 0 & 0 & 0 \\ -1 & 0 & -1 & 0 & 0 & 0 \\ -1 & 0 & 0 & 0 & 0 & 0 \\ 0 & 0 & 0 & 3 & 0 & 1 \\ 0 & 0 & 0 & 0 & 2 & 0 \\ 0 & 0 & 0 & -1 & 0 & 1 \end{pmatrix}$, on $\mathbf{R}^6$.

3 Let $A$ be an $n \times n$ matrix with elements being complex numbers, interpret $A$ as an operator on $\mathbf{C}^n$. Prove that $A$ satisfies its own characteristic equation. This result is known as the *Cayley-Hamilton theorem*. (Hint: If $p(\lambda)$ is the characteristic polynomial of $A$ prove that $p(A) = O$ proving that $p(A)\,X = O$ for arbitrary $X \varepsilon \mathbf{C}^n$. Express $X$ in terms of a basis of generalized eigenvectors.)

## 8.8 Jordan Canonical Forms

Here we analyze the generalized eigenspace $N$ of an operator $L$ corresponding to an eigenvalue $\lambda$ and show how the matrix representation of the restriction of $L$ can be simplified by selecting a suitable basis for $N$. When this is carried out separately for each generalized eigenspace the Jordan canonical form of the operator is obtained according to the theory of the previous section. Let $r$ be the index of $N$. Then $N$ is the kernel of $(L - \lambda I)^r$.

Let $N_j$ $(0 \leqq j \leqq r)$ be the kernel of $(L - \lambda I)^j$ acting on $N$. Hence $N_0$ contains only the zero vector and $N_r = N$.

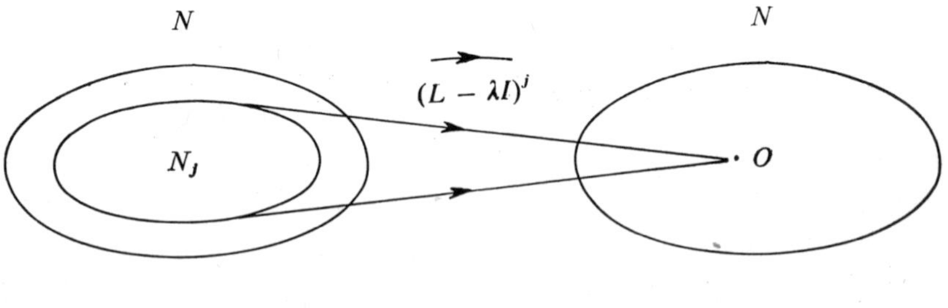


*Lemma 1*

$N_p \leqq N_q$ of $p < q$.

Proof: Let $v \varepsilon N_p$. Hence $(L - \lambda I)^p\, v = O$ implying that $(1 - \lambda I)^q v = O$, giving that $v \varepsilon N_q$, proving the lemma.

Since $N_j$ is the kernel of $(L - \lambda I)^j$ acting on $N$ and since $N_j \subseteq N_{j+1}$ by the lemma, it is seen that $N_j$ is the kernel of the restriction of $(L - \lambda I)^j$

to $N_{j+1}$. Let $P_j$ be a space of progenitors of the range of $(L - \lambda I)^j$ acting on $N_{j+1}$. Then we know from section 3.4 that

$$N_{j+1} = N_j \oplus P_j \qquad 0 \leqq j \leqq r - 1.$$

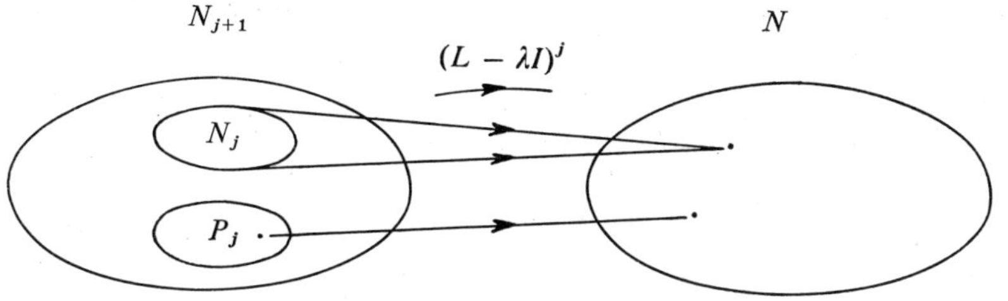


Since $N_r = N$ and $N_0 = \{O\}$, and $N$ is of finite dimension,

$$N = P_{r-1} \oplus P_{r-2} \oplus \cdots \oplus P_0.$$

To find the suitable basis for $N$ that gives a canonical matrix representation of the restriction of $L$ we use this partition of $N$, finding bases in each of the $P_i$. We start with an arbitrary basis for $P_{r-1}$ then show that $(L - \lambda I)$ acting on this set gives a set of linearly independent vectors in a $P_{r-2}$. This set is completed to give a basis for a $P_{r-2}$ and $(L - \lambda I)$ is used again to generate linearly independent vectors in a $P_{r-3}$, etc. The following lemma justifies this procedure.

*Lemma 2*

If the vectors $\{p_i\}$ form a basis for $P_j$, the vectors $\{(L - \lambda I) p_i\}$ are linearly independent vectors in a $P_{j-1}$.
Proof: Consider the identity

$$a_1(L - \lambda I) p_1 \cdots + a_s(L - \lambda I) p_s = O,$$

for arbitrary scalars $a_1, \ldots, a_s$ where $\{p_1, \ldots, p_s\}$ is a basis for $P_j$. This implies that $a_1(L - \lambda I)^j p_1 + \cdots + a_s(L - \lambda I)^j p_s = O$. But the set $\{(L - \lambda I)^j p_i\}$ forms a basis for the range of $(L - \lambda I)^j$, by the definition of $P_j$. Hence $a_1 = \cdots = a_s = 0$ and the vectors $\{(L - \lambda I) p_i\}$ are linearly independent.

It remains to prove that $(L - \lambda I) p_i$ is an element of some $P_{j-1}$. Since $N_j = N_{j-1} \oplus P_{j-1}$ we can do this by proving that $(L - \lambda I) p_i$ is an element of $N_j$ but that it is not an element of $N_{j-1}$. Hence it can be used as a base vector for a space $P_{j-1}$. Since $p_i \varepsilon P_j$, then $p_i \varepsilon N_{j+1}$. Therefore

$$(L - \lambda I)^{j+1} p_i = O,$$

$$\Rightarrow (L - \lambda I)^j (L - \lambda I) p_i = O,$$

$$\Rightarrow (L - \lambda I) p_i \varepsilon N_j.$$

Further, since $p_i \in P_j$, $(L - \lambda I)^j p_i \neq O$,

$$\Rightarrow (L - \lambda I)^{j-1}(L - \lambda I) p_i \neq O,$$

$$\Rightarrow (L - \lambda I) p_i \notin N_{j-1}.$$

Hence $(L - \lambda I) p_i$ can be used as a base vector for a space $P_{j-1}$ and the theorem is proved.

*Construction of the appropriate basis in N.*

We know that
$$N = P_{r-1} \oplus P_{r-2} \oplus \cdots \oplus P_0.$$

Here $P_{r-1}$ is a space of progenitors of $(L - \lambda I)^{j-1}$ acting on $N$. Its elements $\{p_i\}$ will thus be elements of $N$, that is $(L - \lambda I)^r p_i = O$, such that $(L - \lambda I)^{r-1} p_i \neq O$. Hence they will be generalized eigenvectors of rank $r$. Choose $p_1, \ldots, p_s$ to be a basis for such a space $P_{r-1}$. then the vectors $\{(L - \lambda I) p_i\}$ will be generalized eigenvectors of rank $r - 1$ and will be linearly independent in a space $P_{r-2}$. Complete the set by adding the set $\{q_i\}$, the elements of which will be generalized eigenvectors of rank $r - 1$, to give a basis for a space $P_{r-2}$. The set $\{(L - \lambda I) p_i, q_j\}$ will thus be a basis for a space $P_{r-2}$. Further, the set $\{(L - \lambda I)^2 p_i, (L - \lambda I) q_j\}$ will be linearly independent vectors in a $P_{r-3}$. They will be generalized eigenvectors of rank $r - 2$. Complete this set to give a basis for a $P_{r-3}$, etc. Eventually, using this method, since $N$ is a finite dimensional vector space, one constructs in this manner a basis for $N$. These are the base vectors of $N$ that lead to the canonical representation of the restriction of $L$ to $N$. We must further stipulate however the order in which we select these vectors to give the canonical representation.

The above method of selection of base vectors for $N$ makes these vectors elements of chains in $N$. For example corresponding to the base vector $p_i$ of $P_{r-1}$, which is a generalized eigenvector of rank $r$, there is the chain $p_i$, $(L - \lambda I) p_i, \ldots, (L - \lambda I)^{r-1} p_i$ and corresponding to the base vector $q_i$ of $P_{r-2}$ there is the chain $q_i, (L - \lambda I) q_i, \ldots, (L - \lambda I)^{r-2} q_i$. The last vector in each of these chains, as has been stated previously is an eigenvector. The manner of selection of the base vectors is to keep them together in the natural sequence of chains but reversing the order of each chain, starting with the chains of maximal length. For example, let $\{p_1, \ldots, p_s\}$ be a basis for $P_{r-1}$ and $\{(L - \lambda I) p_1, \ldots, (L - \lambda I) p_s, q_1, \ldots, q_h\}$ be a basis for $P_{r-2}$ etc. then we choose the base vectors in the following order:
$(L - \lambda I)^{r-1} p_1, \ldots, (L - \lambda I) p_1, p_1, \ldots, (L - \lambda I)^{r-1} p_s, \ldots, (L - \lambda I) p_s, p_s,$
$(L - \lambda I)^{r-2} q_1, \ldots, (L - \lambda I) q_1 q_1, \ldots, (L - \lambda I)^{r-2} q_h, \ldots, (L - \lambda I) q_h, q_h, \ldots$.

With the base vectors of $N$ selected thus the matrix representation of the restriction of $L$ is readily written down. Let the basis be $\{v_1, \ldots, v_n\}$.

Then $(L - \lambda I) v_j = O$ if $v_j$ is one of the basis vectors which is an eigenvector, $(L - \lambda I) v_j = v_{j-1}$ otherwise.

Hence

$$Lv_j = \lambda v_j$$

or

$$Lv_j = 1v_{j-1} + \lambda v_j.$$

Therefore the matrix of coefficients is of the following form:

$$\begin{bmatrix}
\lambda & 0 & & & & & & & \\
1 & \lambda & 0 & & & & & & \\
 & 1 & \lambda & 0 & & & O & & \\
 & & & \ddots & & & & & \\
 & & & & 0 & \lambda & 0 & & \\
 & & & & & 1 & \lambda & 0 & \\
 & O & & & & & 1 & \lambda & 0 \\
 & & & & & & & & \ddots \\
 & & & & & & & & \lambda
\end{bmatrix}.$$

Here $\lambda$ will be the only element appearing in row $j$, as the diagonal element, if $v_j$ is an eigenvector. It will be preceded in row $j$ by 1 if $v_j$ is not an eigenvector. The transpose of this matrix gives the matrix representation of $L$ relative to this basis. It will be the following type of matrix:

$$\begin{bmatrix}
\lambda & 1 & & & & & & \\
0 & \lambda & 1 & & & & & \\
 & & \ddots & 0 & & & O & \\
 & & & \lambda & 1 & & & \\
 & & & & \lambda & 1 & & \\
O & & & & & \lambda & & \\
 & & & & & & \ddots & \\
 & & & & & & & \lambda
\end{bmatrix}.$$

Here the 1's in the super diagonal will be an indication of the lengths of the chains of generalized eigenvectors.

This matrix is called a Jordan canonical form of the restriction of $L$ to $N$. As explained in the previous section this procedure is carried out for each generalized null space of $L$ to get the *Jordan canonical form* of $L$. Each Jordan block can be represented in the above manner with the appropriate eigenvalue used in each case.

This method involves finding each generalized eigenspace. The following theorem which gives the dimension of each generalized eigenspace will prove useful.

*Theorem 1*

The dimension of the generalized eigenspace $N_i$ of an operator $L$ is the multiplicity of the eigenvalue $\lambda_i$ as a root of the characteristic equation of $L$.

Proof: Let $A$ be a Jordan block form of $L$ and the matrices $A_1, \ldots, A_h$ be the Jordan blocks of $A$. Then the characteristic equation of $L$ is defined to be the characteristic equation of $A$. Hence the eigenvalues of $L$ are the roots of

$$D(A - \lambda I) = 0,$$

where $I$ is the appropriate unit matrix. This equation in $\lambda$ reduces to

$$\pi_i D(A_i - \lambda I_i) = 0,$$

where $\pi_i$ indicates product and $I_i$ is the unit matrix of the same kind as $A_i$. The characteristic equation of $A$ is thus the product of the characteristic equations of the matrices $A_1, \ldots, A_h$. The eigenvalues of $A$ and hence of $L$ are therefore the eigenvalues of the matrices $A_1, \ldots, A_h$. Since $A_i$ is a matrix representation of the restriction of $L$ to the generalized eigenspace of $\lambda_i$ its only eigenvalue is $\lambda_i$. By the fundamental theorem of algebra the number of roots of the characteristic equation of $A_i$ is equal to the order of the characteristic equation. Since the order of this equation is the dimension of the generalized eigenspace $N_i$, and $\lambda_i$ is the only root, the multiplicity of $\lambda_i$ must be equal to the dimension of $N_i$.

In the following examples when one is asked to determine the Jordan canonical form of a matrix $A$ it will be assumed that $A$ is to be interpreted as a linear operator on some $\mathbf{R}^n$.

*Example 1*

Determine a Jordan canonical form of the matrix

$$A = \begin{pmatrix} 3 & 1 & 2 & 0 & 0 & 0 \\ -1 & 0 & -1 & 0 & 0 & 0 \\ -1 & 0 & 0 & 0 & 0 & 0 \\ 0 & 0 & 0 & 3 & 0 & 1 \\ 0 & 0 & 0 & 0 & 2 & 0 \\ 0 & 0 & 0 & -1 & 0 & 1 \end{pmatrix}.$$

(This is the matrix of exercise 2c), section 8.7.) The characteristic equation of $A$ is $(\lambda - 1)^3 (\lambda - 2)^3 = 0$. Hence there are two eigenvalues 1 and 2, both of multiplicity 3. There will be two generalized eigenspaces $N_1$ and $N_2$, and by the previous theorem we know that these will both be of dimension three. It is first necessary to determine these generalized eigenspaces.

The successive powers of $(A - I_6)$ are:

$$(A - I_6) = \begin{pmatrix} 2 & 1 & 2 & 0 & 0 & 0 \\ -1 & -1 & -1 & 0 & 0 & 0 \\ -1 & 0 & -1 & 0 & 0 & 0 \\ 0 & 0 & 0 & 2 & 0 & 1 \\ 0 & 0 & 0 & 0 & 1 & 0 \\ 0 & 0 & 0 & -1 & 0 & 0 \end{pmatrix},$$

$$(A - I_6)^2 = \begin{pmatrix} 1 & 1 & 1 & 0 & 0 & 0 \\ 0 & 0 & 0 & 0 & 0 & 0 \\ -1 & -1 & -1 & 0 & 0 & 0 \\ 0 & 0 & 0 & 3 & 0 & 2 \\ 0 & 0 & 0 & 0 & 1 & 0 \\ 0 & 0 & 0 & -2 & 0 & -1 \end{pmatrix},$$

$$(A - I_6)^3 = \begin{pmatrix} 0 & 0 & 0 & 0 & 0 & 0 \\ 0 & 0 & 0 & 0 & 0 & 0 \\ 0 & 0 & 0 & 0 & 0 & 0 \\ 0 & 0 & 0 & 4 & 0 & 3 \\ 0 & 0 & 0 & 0 & 1 & 0 \\ 0 & 0 & 0 & -3 & 0 & -2 \end{pmatrix}, \text{ etc.}$$

Since $(A - I_6)^3 \begin{pmatrix} a \\ b \\ c \\ 0 \\ 0 \\ 0 \end{pmatrix} = O$, for arbitrary $a, b, c \in \mathbf{R}$ and $N_1$ is three dimensional then $N_1$ must be the space $\{(a, b, c, 0, 0, 0) : a, b, c \in \mathbf{R}\}$. The value of the previous theorem is now seen; in finding $N_i$ one only need look at the kernels of $(A - \lambda_i I_i), (A - \lambda_i I_i)^2 \ldots$ until one reaches the kernel of required dimension. It is then known that further powers of $(A - \lambda_i I_i)$ lead to no new generalized eigenvectors.

The successive powers of $(A - 2I_6)$ are:

$$(A - 2I_6) = \begin{pmatrix} 1 & 2 & 2 & 0 & 0 & 0 \\ -1 & -2 & -1 & 0 & 0 & 0 \\ -1 & 0 & -2 & 0 & 0 & 0 \\ 0 & 0 & 0 & 1 & 0 & 1 \\ 0 & 0 & 0 & 0 & 0 & 0 \\ 0 & 0 & 0 & -1 & 0 & -1 \end{pmatrix},$$

$$(A - 2I_6)^2 = \begin{pmatrix} -2 & -1 & -3 & 0 & 0 & 0 \\ 2 & 3 & 2 & 0 & 0 & 0 \\ 1 & -1 & 2 & 0 & 0 & 0 \\ 0 & 0 & 0 & 0 & 0 & 0 \\ 0 & 0 & 0 & 0 & 0 & 0 \\ 0 & 0 & 0 & 0 & 0 & 0 \end{pmatrix}, \text{ etc.}$$

Since $(A - 2I_6)^2 \begin{pmatrix} 0 \\ 0 \\ 0 \\ p \\ q \\ r \end{pmatrix} = O$, for arbitrary $p, q, r \in \mathbf{R}$ and $N_2$ is a three dimensional space then $N_2$ must be the space $\{(0, 0, 0, p, q, r) : p, q, r \in \mathbf{R}\}$.

It can be seen that $\mathbf{R}^6 = N_1 \oplus N_2$.

We now analyze the generalized eigenspace $N_1$. Since the dimension of $N_1$ is 3 the generalized eigenvector of highest possible rank that could exist would be one of rank 3. This would generate a chain containing further generalized eigenvectors of ranks 2 and one. These could then be used as a basis for $N_1$. Let us investigate to see if generalized eigenvectors of rank 3 do exist. It is seen that for the vector $v = (1, 1, 1, 0, 0, 0)$,

$$(A - I_6)^3\, v = O \quad \text{but} \quad (A - I_6)^2\, v \neq O.$$

Hence $v$ is a generalized eigenvector of rank 3 and it generates the generalized eigenvectors $(A - I_6)\, v$ and $(A - I_6)^2\, v$ of ranks 2 and 1 respectively. Therefore to get the canonical matrix representation of the restriction of $A$ to $N_1$ we use the base vectors $(A - I_6)^2\, v$, $(A - I_6)\, v$, $v$.

$N_2$ is now analyzed in a similar manner. It is seen that any vector $u$ that satisfies $(A - 2I_6)^3\, u = O$ also satisfies $(A - 2I_6)^2\, u = O$. Hence there are no generalized eigenvectors of rank 3 in $N_2$. We now search for generalized eigenvectors of rank 2. It is seen that for $u = (0, 0, 0, 1, 1, 1)$, $(A - 2I_6)^2\, u = 0$ but $(A - 2I_6)\, u \neq O$. Hence $u = (0, 0, 0, 1, 1, 1)$ is a generalized eigenvector of rank 2. It generates a chain which has the additional generalized eigenvector of rank 1, $(A - 2I_6)\, u$. One further vector is needed to form a basis for $N_2$. It will have to be an eigenvector since a generalized eigenvector would generate a chain consisting of two vectors. There is no room for four linearly independent vectors in $N_2$. We find that $w = (0, 0, 0, 0, 1, 0)$ is an eigenvector which is linearly independent of $(A - 2I_6)\, u$. The basis for $N_2$ is taken to be

$$\{(A - 2I_6)\, u, u, w\}.$$

Hence a basis that should lead to a Jordan form of $A$ is

$$\{(A - I_6)^2 v, (A - I_6) v, v, (A - 2I_6) u, u, w\}.$$

These are the vectors $(3, 0, -3, 0, 0, 0)$, $(5, -3, -2, 0, 0, 0)$, $(1, 1, 1, 0, 0, 0)$, $(0, 0, 0, 2, 0, -2)$, $(0, 0, 0, 1, 1, 1)$, $(0, 0, 0, 0, 1, 0)$. The matrix representation of $A$ relative to this basis is found to be

$$\begin{pmatrix} 1 & 1 & 0 & 0 & 0 & 0 \\ 0 & 1 & 1 & 0 & 0 & 0 \\ 0 & 0 & 1 & 0 & 0 & 0 \\ 0 & 0 & 0 & 2 & 1 & 0 \\ 0 & 0 & 0 & 0 & 2 & 0 \\ 0 & 0 & 0 & 0 & 0 & 2 \end{pmatrix}$$

This is a Jordan form of the matrix.

The Jordan form may be interpreted thus, according to the general theory: it has the eigenvalues on the main diagonal, each appearing the number of times of its multiplicity as a root of the characteristic equation. The unities in the super diagonal correspond to the chains in the basis. In the above example there is one chain of length three corresponding to the eigenvalue 1 and two chains of lengths two and one corresponding to the eigenvalue 2. Further examples are now given to illustrate this interpretation.

*Example 2*

Let $A$ be a matrix having two eigenvalues, 1 and 2, of multiplicities four and two respectively. We shall discuss some of the likely Jordan forms.

a) $$\begin{pmatrix} 1 & 1 & 0 & 0 & 0 & 0 \\ 0 & 1 & 1 & 0 & 0 & 0 \\ 0 & 0 & 1 & 1 & 0 & 0 \\ 0 & 0 & 0 & 1 & 0 & 0 \\ 0 & 0 & 0 & 0 & 2 & 1 \\ 0 & 0 & 0 & 0 & 0 & 2 \end{pmatrix}$$

This would be a Jordan form of $A$ if the generalized eigenspace of 1 contained a chain of generalized eigenvectors of length four, and the generalized eigenspace of 2 contained a chain of generalized eigenvectors of length two.

b)
$$\begin{pmatrix} 1 & 1 & 0 & 0 & 0 & 0 \\ 0 & 1 & 1 & 0 & 0 & 0 \\ 0 & 0 & 1 & 0 & 0 & 0 \\ 0 & 0 & 0 & 1 & 0 & 0 \\ 0 & 0 & 0 & 0 & 2 & 0 \\ 0 & 0 & 0 & 0 & 0 & 2 \end{pmatrix}$$

This would be a Jordan form of $A$ if the generalized eigenspace of 1 contained two chains of generalized eigenvectors, of lengths three and one, and the generalized eigenspace of 2 contained two chains, each of length one.

There are further possibilities. Only one of these forms will of course be the correct one, depending on the actual matrix $A$ given.

*Example 3*

Let $A$ be a matrix having the following Jordan form. Interpret the Jordan form to give information concerning the matrix.

$$\begin{pmatrix} 4 & 1 & 0 & 0 & 0 & 0 & 0 \\ 0 & 4 & 0 & 0 & 0 & 0 & 0 \\ 0 & 0 & 4 & 1 & 0 & 0 & 0 \\ 0 & 0 & 0 & 4 & 0 & 0 & 0 \\ 0 & 0 & 0 & 0 & 3 & 1 & 0 \\ 0 & 0 & 0 & 0 & 0 & 3 & 0 \\ 0 & 0 & 0 & 0 & 0 & 0 & 3 \end{pmatrix}$$

The matrix $A$ will have two eigenvalues 4 and 3 of multiplicities four and three respectively. Hence the generalized eigenspace of the eigenvalue 4 is four dimensional. The base vectors of this space are chosen to consist of two chains, each of length two vectors. Hence the eigenspace is two dimensional. The generalized eigenspace of 3 is three dimensional with base vectors consisting of one chain of length two and the other of length one. The eigenspace here is two dimensional. (The last vector of each chain is an eigenvector, these must span the eigenspace.)

*Theorem 2*

A square matrix $A$ over the field of complex numbers can always be transformed into a Jordan canonical form $B$ using a similarity transformation.

The transforming matrix $C$ has columns consisting of generalized eigenvectors of $A$.

Proof: Interpret $A$ as an operator and let $B$ be the Jordan canonical form of $A$ relative to the basis $\{v_1, \ldots, v_n\}$ of generalized eigenvectors. Then $A$ and $B$ being representations of the same operator are related through a similarity transformation, $B = C^{-1}AC$. The matrix $C$ is the matrix representation of the identity operator on the space relative to the usual basis and the basis of generalized eigenvectors (example 2, section 6.4). $C$ has the generalized eigenvectors $v_1, \ldots, v_n$ as columns (exercise 4, section 6.3).

*Example 4*

It can be seen for the matrix discussed in example 1 that

$$\begin{pmatrix} 1 & 1 & 0 & 0 & 0 & 0 \\ 0 & 1 & 1 & 0 & 0 & 0 \\ 0 & 0 & 1 & 0 & 0 & 0 \\ 0 & 0 & 0 & 2 & 1 & 0 \\ 0 & 0 & 0 & 0 & 2 & 0 \\ 0 & 0 & 0 & 0 & 0 & 2 \end{pmatrix} =$$

$$\begin{pmatrix} 3 & 5 & 1 & 0 & 0 & 0 \\ 0 & -3 & 1 & 0 & 0 & 0 \\ -3 & -2 & 1 & 0 & 0 & 0 \\ 0 & 0 & 0 & 2 & 1 & 0 \\ 0 & 0 & 0 & 0 & 1 & 1 \\ 0 & 0 & 0 & -2 & 1 & 0 \end{pmatrix}^{-1} \begin{pmatrix} 3 & 1 & 2 & 0 & 0 & 0 \\ -1 & 0 & -1 & 0 & 0 & 0 \\ -1 & 0 & 0 & 0 & 0 & 0 \\ 0 & 0 & 0 & 3 & 0 & 1 \\ 0 & 0 & 0 & 0 & 2 & 0 \\ 0 & 0 & 0 & -1 & 0 & 1 \end{pmatrix} \begin{pmatrix} 3 & 5 & 1 & 0 & 0 & 0 \\ 0 & -3 & 1 & 0 & 0 & 0 \\ -3 & -2 & 1 & 0 & 0 & 0 \\ 0 & 0 & 0 & 2 & 1 & 0 \\ 0 & 0 & 0 & 0 & 1 & 1 \\ 0 & 0 & 0 & -2 & 1 & 0 \end{pmatrix}$$

One use of the Jordan form of a matrix is in determining solutions to systems of linear differential equations with constant coefficients. If $X'(t) = AX(t)$ is such a system (section 2.2) then a change of variable $X = CY$, where $C$ is some nonsingular constant matrix transforms it into the system $Y' = C^{-1}ACY$. The new system is considerably simpler than the original one if the matrix $B = C^{-1}AC$ is chosen to be the Jordan canonical form of $A$. Hence the transformation matrix $C$ that is used is constructed from the generalized eigenvectors of $A$. This method is primarily of theoretical value.[1,2]

---

[1] For a further discussion of this method the reader is referred to F. Brauer and J. A. Nobel, Qualitative Theory of Ordinary Differential Equations, section 2–6 (W. A. Benjamin Inc., 1969).

[2] For a further discussion of the spectral theory of operators see, for example, B. Friedman, Principles and Techniques of Applied Mathematics (J. Wiley and Sons, Inc. 1965).

## Exercises

1 Determine the Jordan canonical forms of the matrices

$$\text{a)}\ \begin{pmatrix} 2 & 0 & 4 & 3 \\ 0 & 2 & 0 & 2 \\ 0 & 0 & 1 & 1 \\ 0 & 0 & 0 & 1 \end{pmatrix} \qquad \text{b)}\ \begin{pmatrix} 4 & 3 & 2 & 4 & 1 \\ 0 & 4 & 1 & 0 & 1 \\ 0 & 0 & 4 & 2 & 2 \\ 0 & 0 & 0 & 1 & 3 \\ 0 & 0 & 0 & 0 & 1 \end{pmatrix}$$

that were discussed in the exercises of the previous section. Illustrate that these forms can be obtained using similarity transformations.

2 Let $A$ be a square matrix having three eigenvalues 1, 2 and 3 of multiplicities three, two and one respectively. Give all possible Jordan forms of $A$ discussing the properties of $A$ (as in example 3) for each case.

3 Let $A$ be a square matrix having two eigenvalues 4 and 5, both of multiplicity two. Give all possible Jordan forms of $A$ discussing the properties of $A$ (as in example 3) for each case.

## Bibliography

Brauer, F., and Nohel, J. A., *Qualitative Theory of Ordinary Differential Equations* (W. A. Benjamin Inc., 1969).

Dantzig, George B., *Linear Programming and Extensions* (Princeton University Press, 1963).

Faddeev, D. K., and Faddeeva, V. N., *Computational Methods of Linear Algebra* (W. H. Freeman and Company, 1963).

Friedman, B., *Principles and Techniques of Applied Mathematics* (J. Wiley and Sons, Inc., 1965).

Kreider, D. L., Kuller, R. G., and Ostberg, D. R., *Elementary Differential Equations* (Addison-Wesley, 1969).

Luenberger, D. G., *Optimization by Vector Space Methods* (J. Wiley and Sons, Inc., 1969).

Nobel, Ben, *Applied Linear Algebra* (Prentice Hall, 1969).

Simmons, G. F., *Topology and Modern Analysis* (McGraw-Hill, 1963).

Yosida, K., *Functional Analysis* (Springer, 1965).

# Index